SAGE was founded in 1965 by Sara Miller McCune to support the dissemination of usable knowledge by publishing innovative and high-quality research and teaching content. Today, we publish more than 850 journals, including those of more than 300 learned societies, more than 800 new books per year, and a growing range of library products including archives, data, case studies, reports, conference highlights, and video. SAGE remains majority-owned by our founder, and after Sara's lifetime will become owned by a charitable trust that secures our continued independence.

Los Angeles | London | New Delhi | Singapore | Washington DC

Assorted

Assorted

CITY

Equity, Justice, and Politics in Urban Services Delivery

SUPTENDU P. BISWAS

SAGE www.sagepublications.com
Los Angeles • London • New Delhi • Singapore • Washington DC

First published in 2015 by

SAGE Publications India Pvt Ltd
B1/I-1 Mohan Cooperative Industrial Area
Mathura Road, New Delhi 110 044, India
www.sagepub.in

SAGE Publications Inc
2455 Teller Road
Thousand Oaks, California 91320, USA

SAGE Publications Ltd
1 Oliver's Yard, 55 City Road
London EC1Y 1SP, United Kingdom

SAGE Publications Asia-Pacific Pte Ltd
3 Church Street
#10-04 Samsung Hub
Singapore 049483

Published by Vivek Mehra for SAGE Publications India Pvt Ltd, typeset at 10/12.5 pts Times New Roman by Emaptis, Chennai, and printed at Sai Print-o-pack New Delhi.

Library of Congress Cataloging-in-Publication Data

Biswas, Suptendu P.
Assorted city : equity, justice, and politics in urban services delivery / Suptendu P. Biswas.
pages cm
    Includes bibliographical references and index.
    1. Urban policy—India. 2. City planning—India. 3. Municipal services—India—Planning. I. Title.

HT147.I4B57   307.1'2160954—dc23   2015   2015015908

ISBN: 978-93-515-0125-1 (HB)

The SAGE Team: Shambhu Sahu, Neha Sharma, Megha Dabral, and Rajib Chatterjee

To

Late Professor Souro D. Joardar and all my teachers

Bulk Sales

SAGE India offers special discounts
for purchase of books in bulk.
We also make available special imprints
and excerpts from our books on demand.

For orders and enquiries, write to us at

Marketing Department
SAGE Publications India Pvt Ltd
B1/I-1, Mohan Cooperative Industrial Area
Mathura Road, Post Bag 7
New Delhi 110044, India

E-mail us at **marketing@sagepub.in**

Get to know more about SAGE

Be invited to SAGE events, get on our mailing list.
Write today to **marketing@sagepub.in**

This book is also available as an e-book.

Contents

List of Tables

List of Figures

List of Abbreviations

AC	Assembly Constituency
ANOVA	Analysis of Variance
CDP	City Development Plan
CPHEEO	Central Public Health and Environmental Engineering Organization
DCB	Delhi Cantonment Board
DDA	Delhi Development Authority
DJB	Delhi Jal Board
DMC	Delhi Municipal Corporation
DUIIP	Delhi Urban Environment and Infrastructure Improvement Project
DWSP	Delhi Water Supply and Sewerage Project
EIUS	Environmental Improvement in Urban Slums
FDS	Factor of Duration of Supply
FWS	Factor of Water Supply
GNCTD	Government of National Capital Territory of Delhi
HAC	Hardship and Anomalies Committee
HIG	Higher Income Group
HUDA	Haryana Urban Development Authority
IPG	Ideology–Policy–Governmentality
ISA	Ideological State Apparatus
JJ	*Jhuggi Jhopri*
JNU	Jawaharlal Nehru University
JNNURM	Jawaharlal Nehru National Urban Renewal Mission
LIG	Lower Income Group
LoS	Level of Satisfaction
lpcd	Liter per Capita per Day
MCD	Municipal Corporation of Delhi
MGD	Million Gallons per Day
MIG	Middle Income Group
MLA	Member of Legislative Assembly
MLALAD	Member of Legislative Assembly Local Area Development
MoUD	Ministry of Urban Development

MPD	Master Plan of Delhi
MPLAD	Member of Parliament Local Area Development
MW	Municipal Ward
NCRPB	National Capital Region Planning Board
NCT	National Capital Territory
NDMC	New Delhi Municipal Council (earlier, New Delhi Municipal Committee)
NIUA	National Institute of Urban Affairs
NGO	Non-Governmental Organization
NRW	Non-Revenue Water
NSSO	National Sample Survey Organization
OECD	Organization of Economic Co-operations and Development
OZ	Operational Zones
PC	Parliament Constituency
PPP	Public–Private Partnership
RSA	Repressive State Apparatus
S&JJ	Slums and *Jhuggi Jhopri*
SPA	School of Planning and Architecture
TCMD	Thousand Cubic Meters per Day
UAV	Unit Area Value
UFW	Unaccounted for Water
WtP	Willingness to Pay
WTP	Water Treatment Plant

Preface

If I look back, I will find a few day-to-day observations, casual conversations, and certain newspaper reports from where this work began.

Nehru Camp, Navjeevan Camp, and Bhoomiheen Camp are three adjacent squatter settlements forming one of the largest clusters in South Delhi next to the housing pockets at Kalkaji Extension, where I have been living for almost a decade now. Slums or squatter settlements like these are not provided with several basic needs including piped water supply. Delhi Jal Board (DJB) makes arrangement for the water tankers to supply water to these areas (Figure P.1). Whenever a tanker comes to these areas, it creates a huge ruckus among the residents of these camps. Everyone rushes toward the tanker to get the first drop of water; few of them manage to climb up the tanker to reach the water source; some quarrel about their queue positions! The vehicular road eventually gets blocked as tankers are parked there. It is, indeed, sad to see people struggling so much for a basic service, like water supply, in the capital city of India. But, in our colony situated next to these slums, my neighbor's water tank is always seen overflowing indicating the other state of existence and its callous manifestations exposing instances of localized disparity.

Figure P.1:
Water Supply by DJB Tankers

Source: Vina Biswas (2011).

A road near our house had been full of potholes for more than five years before it got repaired on the eve of an election. The particular road was the boundary of two adjacent assembly constituencies (ACs) before the delimitation happened. Local people, in general, used to blame the lack of *political responsibilities* for the poor condition of the road. Popular opinion, many a time, tends to hold political initiatives accountable for a work to be attended by the civic authority.

In the last two decades of my stay in New Delhi, I resided in many housing colonies in the southern part of the city. But we have never shifted to Vasant Kunj, a housing scheme promoted by the Delhi Development Authority (DDA) in the late 1980s, especially after we heard how some of our friends staying there had to keep track of supplied water. Localities in and around this part of Delhi clearly had less provision of water and became an example of the territorial variation of water supply across the capital. Nowadays, the supply from the Sonia Vihar water treatment plant has reportedly improved the water situation in Vasant Kunj.

There used to be a squatter settlement opposite Jahanpanah Club in the *posh* neighborhood of Greater Kailash II and Alaknanda. Like many other similar squatters, this one, too, *encroached* upon a vacant DDA-owned land designated for *community facilities*. It had a typical *slum-like* living condition with semi-permanent houses and shops. I remember getting a mattress stitched from one of those shops. Apparently, residents of the neighboring colonies were objecting to the presence of the settlement for some time. Finally, the slum got removed much before the "Sealing Drive" ordered by the High Court of Delhi in 2006. Interestingly, Nehru Camp, Navjeevan Camp, and Bhoomiheen Camp, located within a distance of a kilometer from the removed squatter settlement, continue to exist despite objections from the surrounding low-profile housing colonies. One tends to believe that the slums are arguably removed from places having greater visibility due to the location or the influence of affluent residents or the political access. The absence of a "collective identity or voice" makes the urban poor, in spite of their large population, highly vulnerable to the strategies of the State (Baviskar, 2006). The recent slum clearances by the government on the pretext of the Commonwealth Games have also let this apprehension surface in some writings (ibid.).

Many of us might have sensed similar incidents of *selective* delivery of urban governance across colonies and localities in Delhi. Multiple conditions of selectivity may have many versions of explanation. The removal of the slum near Alaknanda appears to be an illustration of the state favoring specific social groups over others. The implicit involvement of the state as a cold onlooker in other instances can also be felt. Resource constraints in the Vasant Kunj example could not have been overlooked otherwise. Tanker supply in slums is required because of the absence of the fixed installation of network. Slums, too, have grown out of the necessities by subverting the planning policy of the State.

To consider politics being completely oblivious to such exceptions would be naive.

Acknowledgments

A lot of good wishes contributed to the making of this book.

The book was written over two years based on about seven years of research. My doctoral research behind this book would not have been possible without the encouragement and advice of Late Souro D. Joardar of School of Planning and Architecture (SPA), New Delhi and Atiya Habeeb Kidwai of Jawaharlal Nehru University (JNU), New Delhi. Even in my last telephonic conversation with Dr Joardar, before his untimely demise, he was very keen to find out the progress of this book.

I am also thankful to Mahavir and N. Sridharan, both Professors of Planning at SPA, New Delhi, who have always addressed my problems and advised the best possible alternative. Neelima Risbud, former Dean of Studies and Raman D. Surie, former Head of the Department of Physical Planning, both from SPA, New Delhi, helped me in difficult situations during my research.

I also take this opportunity to express my gratitude to Ram Sharma, K.T. Ravindran, Late M.M. Rana, Ashish Choudhury, Ashish Bhalla, Vishal Aggarwal, and many others who, at different times, advised me to write down my thoughts. I am thankful to my students and fellow faculty members for giving me opportunities for meaningful discussions with them on related fields of study.

I owe special thanks to Tejamoy Ghosh, Rajat Ray, and Leon Morenas. Tejamoy has helped me immensely in recapitulating some of the statistical knowledge and introduced me to more advanced programs and statistical methods. Rajat Ray has listened to my moments of disappointment and achievement all the while. I have borrowed many useful books from Leon, who was always willing to share an extra bit of information. Prashant Kumar and Aniket Vishwakarma helped me in redrawing some of the maps. Saurabh Tewari took time out to design the cover.

I am thankful to the entire SAGE Team, especially Vivek Mehra, Shambhu Sahu, and Neha Sharma. Vivek Mehra helped in ironing out contractual issues with his prompt responses. Shambhu Sahu did all possible coordination meticulously and was really instrumental in pursuing me in the process of publication of this book. Neha Sharma coordinated the editorial work to tie all possible loose ends.

Maa, my mother, has always enquired about the progress of my work and encouraged me whenever she got the chance; so have my in-laws: Papa, Mummy, Reena, and Sukanya. My brother, Subhadip, too, is eagerly waiting for the completion of this work.

I am extremely thankful to my wife, Vina, for single-handedly managing all the family matters and accommodating my never-ending engagement with this work. My son, Devneil, has always been appreciative and never complained when I could not give him sufficient attention. He has been waiting to see this book published. Lastly, I remember *Yaa,* my late father, and hope he can see it complete from wherever he is...

At one point, I was unable to sit at the computer because of severe ill-health. I thank the Almighty to enable me to complete this book.

1

Introduction

This book problematizes and attempts at a conceptual reading of the city. *Assorted City* is a notion proposed here for a possible reading of the contemporary city created out of incompatible and similar enclaves. These enclaves are not necessarily terrible spatial segregations but practices of scattered collection of different forms of urban living, often imbued with different conditions of inequity. Enclaves are produced by exceptions. These are exceptions to the *main* parts of the city, to the mainstream policy, to the *planned* city, or to the dominant spaces and alike.

Methodologically, I witness conditions of an assorted city while understanding equity and justice in the delivery of urban water supply. Empirical observations reveal that multiple shades of equity and justice, which I recognize as an *equity mosaic*, exist in a city in place of prevalent binary oppositions (of rich–poor, legal–illegal, planned–unplanned, etc.). Such a mosaic, indeed, gives rise to various situations of enclaves. Consequently, exceptions favoring certain socioeconomic groups and their spaces yield to the mixed condition of (in)equity and (in)justice. In turn, the *politics of distribution* is instigated outside the policy regime. It is important to note that discussions on issues of equity and justice, often, describe conditions of inequity and injustice.

The context of this book is the post-independence planning of Delhi. Independent India envisioned the *idea* of democracy with equality and liberty as its basic tenets. With an *intention* of attaining such an idea, planning was adopted as the policy to take the new nation forward. Delhi was the microcosm and the model of demonstration of similar intentions, but today, its *real* existence is full of multiple conditions of urban living.

Only a few parts of the city somewhat resemble what was originally intended, while the rest has grown outside and beyond the dictum of planning. These are the heterotopias this work is interested in.

It is widely assumed that planning as a policy would adhere to the ideological commitment of equity and justice in the distribution of resources and services in a democracy. However, this work puts forth the argument that planning has in it inherent traits of multiple techniques of governmentality, which necessitates the politics of distribution. Politics of distribution with its unpredictable consequences, then, supersedes the policy of distribution.

In Delhi, I observe this phenomenon in the distribution and delivery of water supply. It is well-established that the urban poor living in squatters get lower water supply from the state agency, namely the Delhi Jal Board (DJB), as compared to the colonies with legal property rights of which planned colonies are also a part. Empirical observations reveal that even across *legal* colonies, multiple delivery conditions exist despite same norms of delivery. In fact, the higher strata of society and the socioeconomic spaces they occupy tend to get more water supply. I recognize the existence of an *equity mosaic* out of multiple shades of equity and justice in a city instead of prevalent binary oppositions (of rich–poor, legal–illegal, planned–unplanned, etc.). Thus, an assorted city, the condition of contemporary urban living, comes into being.

Key questions are: When intended conditions drive policy decisions to deliver what is imagined as just, do intentions compromise with the *ideology of the just*? Once real distributions fall short of what is intended, it is also important to know—how are these shortfalls negotiated in the practices of *governmentality*? Are scarce resources distributed evenly across spaces and people or is there any *spatial selectivity*? Such queries are central to the concern for the idea, intention, and delivery of equity and justice.

In response to these questions, the work outlines three broad objectives: first, to construct and assemble a theoretical framework to capture different ideological moments of the state and related instances of its practices; second, to build around such a framework, an account of justice and equity in the delivery of urban basic services of water supply; and third, to explore possible tactical and systemic nuances in approaching conditions of equity and justice.

The transition of India from the welfare-state condition to the neoliberal one and corresponding changing notions of equity and justice are captured through the ideology–policy–governmentality (IPG) framework within tactical theoretical alliances between the Marxist and the poststructuralist positions. Such framework helps to adjoin and, therefore, explains the idea, intention, and real conditions while moving across different time-spans of ideologies India has adhered to. The mainstream Marxist theories are critically aware of the unevenness and differential conditions of social justice in a historical and geographical sense, whereas the poststructuralist notions describe the ontology of multiple existences and practices.[1] The role of the state and the way governance conducts itself are crucial and central to such a theoretical alliance.

The first of my three central theoretical references for the IPG construct comes from Louis Althusser's readings (1970) of Marx's *Capital* and his essays on "Ideology and Ideological state Apparatuses (Notes toward an Investigation)" (in Zizek 1994). Ideology is the system or the construction of the representations or the ideas having material existence in "an apparatus, and its practice, or practices," which Althusser identifies as the "Ideological State Apparatuses" and the "Repressive State Apparatuses" (ibid.).

The second major work, I refer to, is Jacques Rancière's *Disagreement* (1999). Rancière's central thesis is built upon the notion that the politics of democracy has "the rationality of disagreement," whereas the policy (or police) is the consensual practice of the government to the "conceptual legitimization of a democracy" and, hence, becomes the condition of the "post-democracy" (ibid.: xii, 102).

The third significant work, perhaps the one most used here, is Michel Foucault's discourses on "governmentality" (1979, 1984) as well as on the disciplinary techniques in *Discipline and Punish* (1975[1972], 1984), and other subsequent literature (Barnett et al. 2008, Burchell et al. 1991, Chatterjee 2004, Corbridge et al. 2005, Curtis 2002, Dean 1999, Donzelot and Gordon 2008, Larner 2000, Legg 2005, 2007, Lemke 2001, 2002, 2007, Rose 1999). Governmentality is widely referred to as the "rationalities" or the "mentalities" of the government, performed by multiple agencies "[with] unpredictable consequences, effects and outcomes" (Dean 1999: 18, Miller and Rose 2008).

There would be an apprehension in putting together such a framework from diverse, often contradictory, theoretical sources.[2] On quite a few occasions, poststructuralist discourses of governmentality, for example, are seen antithetical to the Marxist project of ideology (Donzelot and Gordon 2008, Harvey 1996, Lemke 2001, Zizek 1994). Yet, one may see a point of convergence in Balibar's (1992) work on Foucault and Marx and also in Althusser's notions of Ideological State Apparatus (ISA) and Foucault's disciplinary techniques in governmentality.[3]

The question is how does one expect the concepts of ideology, policy, and governmentality to interact in this work? Ideology imparts the imagination around which policy is formed. But most of the time, a modified version of the original policy is implemented. This modification happens due to governmentality, which, consequently, is shaped or influenced by the realities. Realities, then, can alter the world view and, in turn, may change the ideology too. The cycle between the modified policy, governmentality, and realities causes the *politics of distribution* (Figure 1.1). In the generic theoretical scaffolding of IPG, the notion of policy is the central pivot, which can be, on the one hand, approached from the ideological end of the policy-making and, on the other, modified from the governmentality ends of delivery and implementation tactics (Figure 1.1).[4] Politics of distribution, if one has to rely on Rancière's notion

Figure 1.1:
Relationship between Ideology, Policy, and Governmentality

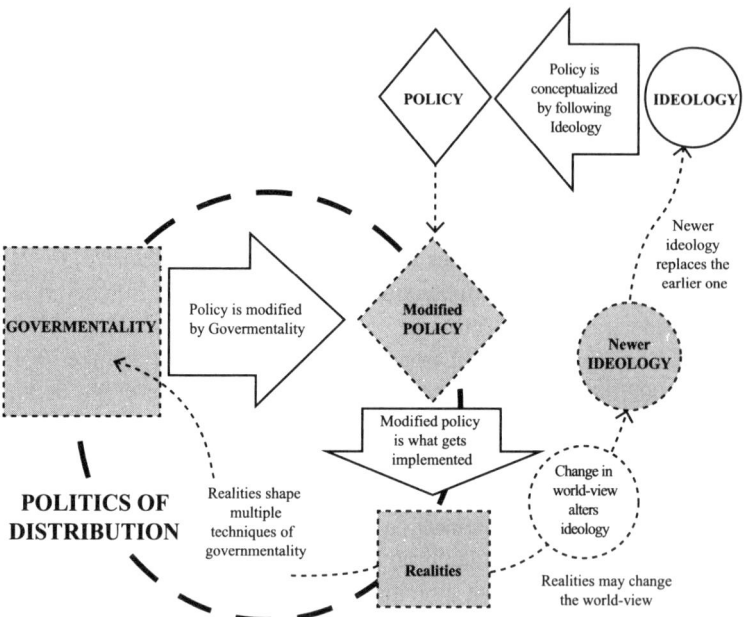

Source: Author.
Note: Ideology forms the imagination around which a policy is formed. However, a modified version of the original policy is what often gets implemented. Governmentality, influenced by the realities, shapes this modification. The cycle formed between the modified policy, governmentality, and realities causes the politics of distribution. Realities, then, can alter the world-view and, in turn, the ideology.

of "disagreement" (1999), essentially represents the divergence from the policy of distribution.[5] And multiple techniques of governmentality, conducting the self and the other formally and informally, contribute to and, at the same time, are outcomes of the politics of distribution.

Now, one attempts to position the discussions on the threefold distributive conditions of resources within the IPG framework. Intended distribution shall be the rational outcome of what is initially considered as the just distribution within the ideological dimensions. For example, a democratic welfare state would have a different understanding of the notion of equity and justice from a military government. In an ideal condition, the intended and the just distribution should be the same (Figure 1.2). But gaps do exist in the actual practice between the just and intended distributions as well as between intended and real distributions. The just, intended, and real distributions indicate respective conditions of existence, namely the *imagined* ideological condition, the *consensual* policy arrangement, and the governmentality of *unpredictable* delivery practices (Althusser 1970, Dean 1999, Foucault 1984, Rancière 1999).

Figure 1.2:
Just, Intended, and Real Distribution: A Conceptual Relationship

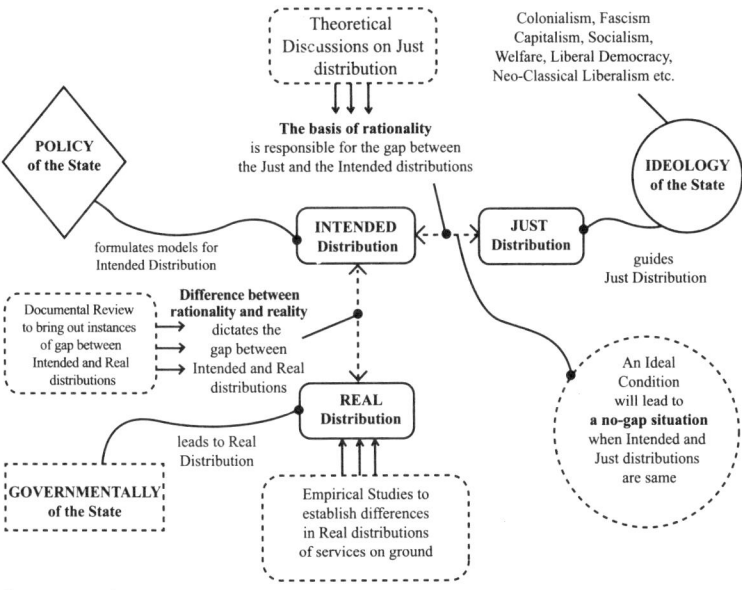

Source: Author.

The IPG framework, thus, would help in explaining the idea, intention, and real conditions of distribution while moving across ideological transitions in different time-spans, for example, the welfare-state condition to the neoliberal one in the independent India (Figure 1.2). Any change in the ideological position is expected to incur corresponding modifications in the notion of equity and justice as well.

It is also relevant to discuss the concept of justice in the context of democracy, ideology of which, on the one hand, is founded upon the ideals of equality and liberty and, on the other, can be recognized as a formal exercise of public reason by many (Fainstein 2005, 2010, Habermas 1995, Rawls 1971, Sen 2009, Young 2011[1990]).[6] Democracy, the convergence of "political participation, dialogue and public interaction," offers a setting where opportunities and choices seem equal, at least conceptually, and can enhance "social justice and a better and fairer politics" (Sen 2009: 326, 351). To reiterate, practice of such an ideology by the state makes a choice of its policy, which is likely to differ from the policy of the state under other ideological conditions.[7]

John Rawls' *A Theory of Justice* (1971) and Amartya Sen's *The Idea of Justice* (2009) are two significant references for me. What makes Sen's capability approach different from Rawls' is its focus on the "*actual opportunities* of living" in place of the "*means* of living" (ibid.: 233, emphasis in original). Opportunities for the enhancement of capabilities are, nevertheless, dependent on the provision or access to means (in terms of services and facilities) and if this need is not addressed in the public provision by the state, it leads to deprivation due to inequality. Both for Rawls (1971) and Sen (2009: 62), the idea of fairness appears "central to justice."

Under such a condition, if the formulation and the implementation of policies have room for any *selectivity* in providing the means or the opportunities, the notion of equity in the distribution and access of resources and services in a city will certainly be shaken; in turn, weaknesses in social justice shall be exposed (Fauconnier 1999, Harvey 1975[1973], Rawls 1971, Runciman 1966). Such selectivity, I argue here, occurs with respect to the socioeconomic groups or the spaces they occupy.

The context of this work is the democratic space of independent India. Ideology of the welfare state of the independent India guided the nation toward "some of the fundamentals of the socialist structure" with the mission to reduce the gap between the rich and the poor to a minimum (Barr 1992, Nehru 1994[1946]: 400, 397). In doing so, policies were framed with an intention to ensure social well-being by the state provision of goods and services in the line of "Keynesian welfarism" (Nehru 1946[1994]). Amidst many flaws, "the ideological commitment" of the Nehruvian State "to social reform and distributive justice" has been believed to be its strength (Kaviraj 2005: 13).[8] For Nehru, "equitable distribution" of resources (food, income, housing, health, education, utilities, etc.) was what would ensure "national self-sufficiency" (Nehru 1994[1946]: 396–398). He conceptualized comprehensive planning as the key method in achieving this goal:[9]

> It was obvious also that any comprehensive planning can take place under a free national government, strong enough and popular enough to be in a position to introduce fundamental changes in the social and economic structure.... Planning thus was not so much for the present, as for *an unascertained future*, and there was *an air of unreality* about it. (ibid., italics included)

Planning was, thus, the policy approached from the ideological ends of just distribution.

The urban context of the master-planned Delhi is where the empirical component of the research is located. Having taken clues from generic ideals of modern planning, Master Plan of Delhi (MPD) used two broad distribution techniques to "provide" for resources over spaces and among people: "decentralization" to achieve self-contained planned units, and "hierarchy" to allocate resources and facilities over planned spaces (DDA 1962). The rationality of distribution that had been conceived by the planning as a part of the "policy of the state" was obstructed by the non-execution and the partial execution of the MPD 1962 and the related bad governance.

The emerging question is: How urban planning, while being implemented, tends to pick up the normative techniques of governmentality in distribution?

I observe the *norms of practice* of the post-independence urban planning by taking the example of the MPD. DDA was set up as per the Delhi Development Act 1957 to formulate the Master Plan with a view to *rationally* control the urban growth through comprehensive planned development. Till date, DDA has prepared three key documents: MPD 1962, MPD Perspective 2001 (MPD 2001), and MPD 2021.

Master Plan, on the one hand, was the production (or the apparatus) of the rationality of the policy of planning, approached from the ideology of the welfare democracy, and, on the other hand, was the *ensemble* delivered through the techniques of governmentality in correspondence with the mentalities (or the rationalities) of the policy of planning.

Along the line of Foucault's ontological account, *rationalities of governmentality* are understood as the ways of thinking or the reasoning of the government behind responding to a problem, whereas *mentalities of the government* choose the way of thinking for its (government's) practices (Dean 1999: 24–25, Foucault 1984). Hence, governmentality tends to apply multiple techniques of normalization and has ingrained in it tendencies of "unpredictable consequences, effects, and outcomes" (Dean 1999: 18, 17–23). Politics of distribution, then, emerges, which essentially differs from the policy of distribution (Table 1.1).

When the governmentality discourse is extended a step further, a trace of spatial selectivity can be seen embedded in the practices (the ways of governance), and mentalities of the government behind such practices may also employ *selective* methods yielding to *uneven* results (Dean 1999, Jessop 2007, Roy 2009, Swyngedouw 2005). Now the question is: How does the selectivity come about in the distribution and delivery of urban resources and services? More importantly, does such delivery get affected by spaces inhabited by groups of people with different income and social status (which I rather loosely term as "socioeconomic space")?

Arguments then return to some of the issues raised earlier: When aspects of lived conditions were thought of or planned, why are there so

Table 1.1:
Differences between the Policy and the Politics of Distribution in a Democratic State

Policy of distribution	• is formulated by the state to ensure equality and justice across its citizens • is a regime, consensually arrived at, to be followed by the state machinery and to be adhered to by citizens • anticipates the uncompromised delivery of the policy • imagines uniform application of the essence of the policy even in distributing scarce resources and services
Politics of distribution	• is a deviation from the policy regime • happens either by the top-down decision-making or by the divergent ground realities at the bottom. • occurs frequently out of multiple delivery strategies undertaken by the state. • is the manipulation of the policy by different interest groups. • is often erratic and unpredictable subversion of the policy "out of necessities"

Source: Author.

many deviations from the stated objectives of the Plan? How much of the Plan has been delivered, where, and to whom?

A tactical theoretical alliance of different positions is attempted to observe how *politics of distribution* out of the multiple techniques of governmentality causes injustice and inequity. I tend to draw references from four broad explanations. First is selective distribution by the state; when a state, an ensemble of institutions, makes decisions on its own, it leads to *selective delivery* in the end (Brenner 2004, Jessop 1990, Jessop et al. 2008, Park 2008).[10] Such selectivity of the state usually upsets its policy guidelines and can be seen as a technique of governmentality. Second is location-specific distributive decisions; this creates inequity and unevenness in accessing resources by favoring "the influential groups" and their occupied spaces (Harvey 1975[1973], Lefebvre 1991[1974], Smith 2008[1984]). Third is growth-centric distributive decisions which are likely to manipulate the distribution and delivery decisions on the basis of the local level "growth coalitions" (Molotch 1976, 1993). Last is the state of exception which is formed when the state (the sovereign), out of necessities, tends to selectively dilute certain structural relations between the law and the society while allowing same relations intact for others; hence, the subjective application of the otherwise objective legal framework is seen maneuvered within the ambiguous margin between the legal and the political in a "state of exception" (Agamben 2005[2003]).

All these arguments explain different conditions of politics of distribution causing inequity and injustice. In case urban planning is controlled by the state, differences between the *intended city* (projected by the policy of urban planning) and the *real city* on ground may be seen in light of the strategic selectivity of the state, further allowed by the inherent unpredictability of the techniques of governmentality. These traits of unpredictability and strategic selectivity in governmentality, which make the *real city* deviate from the *intended city*, have contextual, yet regular occurrence. Similarly in the provision of urban services too, the intended level of supply that the planning policy anticipates may not get implemented or delivered in reality out of the politics of distribution. The unitary notion of the consensus of policy can also be seen disturbed by the *politics of distribution* which tends to generate different conditions of selectivity, both spatial and socioeconomic, which is further connected with the legal selectivity in a state of exception.[11] It is obvious that the ideology of democracy did not have these notions of difference that the politics of democracy has (Chatterjee 1994, 2004, Rancière 1999,

Swyngedouw 2009). Instead, the ideology had the idea of equality and liberty contributing to the construct of the *idea city*.

Delivery of urban water supply in Delhi is used as the key empirical object to understand the equitable distribution within such a theoretical framework (Figure 1.3). The provision of water supply under the control of the state is expected to address some of its ideological commitments. Yet, the narratives of dual urban existences (within the *two-city* notions), for example, the planned–unplanned, legal–illegal, authorized–unauthorized, seem to have implicated the delivery of services in Delhi. These binary constructs of the city are

Figure 1.3:
DDA Planning Divisions (Zones)

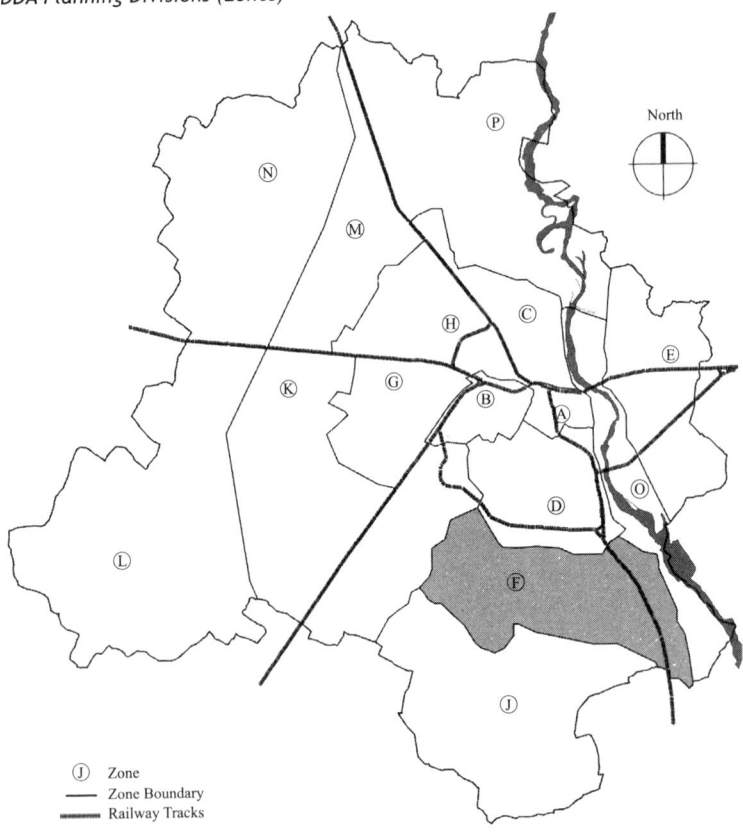

Source: Adapted from DDA (2007a).

not the only ones to which the inequity and injustice are confined. I make an attempt to demonstrate here that even within the so-called planned and legal spaces in Delhi, the delivery of water supply is different across spaces, and higher socioeconomic groups tend to get better supply than the lower ones. Hence, the inequity and injustice in cities seem to have multiple gradations.

Key questions addressed are threefold. The first, how does the delivery of the urban water supply in Delhi correlate with city's socioeconomic spaces and its legal property rights? The second, how does political patronage relate to the socioeconomic spaces (indicated by the property tax strata) in Delhi? And the third, what is the correspondence between legal property rights and socioeconomic spaces?

Water is a scarce resource in Delhi. Thus, the concern for equity in its delivery and distribution across spaces (spatial) and people (socioeconomic) is expected to be covered in the provision by the state. Any trace of socioeconomic and spatial selectivity in water supply would then bring out inequity in real distributions, in contrast to the intended policy (of planning). In turn, conditions of inequity and injustice in accessing urban infrastructure would be created.

Majority of the prevalent discourses revolve around the provision of basic services to the *urban poor*, and/or to the *illegal* occupants of the land in the eyes of the state and the judiciary. However, multiple shades of inequity are revealed in this work across Master-Plan-designated *planned* (and hence, *legal*) colonies having same land tenure status and same targets of water supply, as per the MPD. So, the policy of allocation bears an inherent inequity, compounded further by the governmentality of multiple practices of actual delivery. In fact, one may also feel that the ideology of equality never transpired in the policy formation of the provision of water to the people in the city of Delhi, which instead seems to be confounded with the governmentality decisions.

It is often believed that the distribution of resources (to whom, where, and how much) is manipulated by the local political patronage, in *coalition* with the other influential and powerful residents (Harvey 1975[1973], Lasswell 1936, Logan and Swanstrom 2005[1990], Mollenkopf 1992, Molotch 1976, 1993). On the flip side, the *patron–client relationships* are formed between the politicians (and other influential groups) and the *political society*, comprising of the urban poor living outside the general domain of law and regulation (Chakrabarty 1989, Chatterjee 2004).[12] But these relationships, primarily based on the vote-bank-dominated political strategies, work in favor of the under-privileged by accommodating them within some kind of *para-legal* arrangements (Chatterjee 2004). Both these versions of patronages, produced through many unseen negotiations, can influence the implementation in favor of the richer or poorer sections of the society.

Another important consideration is to examine the correspondence between legal property rights and socioeconomic spaces. Many *two-city* notions understand city as an entity of binary opposites of legal and illegal existences. The idea, here, is to observe the legal notion of property to expose multiple gradations (or shades) of equity and justice across different socioeconomic strata even within the seemingly monolith construct of the *legal* city. I use the property tax categories by the Municipal Corporation of Delhi (MCD) as a tool of the legal property and observe its relationship with the diverse socioeconomic spaces that characterize the residential milieu of Delhi. In the absence of any direct measure of the socioeconomic characteristics, property tax would eventually become a fitting surrogate to classify socioeconomic variations through statistical analysis of the empirical data in order to study the delivery of water supply and the political funding, both, across the planned and/or legal colonies.

Research methods for the empirical work include case study, spatial mapping, stratified random sample survey, reconnaissance survey, review of documents, and statistical analysis. Spatial mapping techniques are used for the delineation of the area within which the clusters of colonies, based on the maps of concentrations of tax strata, were selected for the household survey. A residential colony is considered as the unit of study here. The case study method has been applied for comparing across the units of study. Datasets were compiled from the primary household survey, the review of the documents of the MCD, and that of the DJB for the information on the colony-wise property tax categories, water supply, and the spending of the Member of Legislative Assembly Local Area Development (MLALAD) funds. The reconnaissance survey, too, was carried out to observe and understand ground situations. Statistical analysis, however, has been the most significant method used in this work. Statistics has been a major tool of the policy and the analytics of governmentality. The rationality of planning in Delhi, too, followed statistical techniques.

The application of statistically dependent methods seems to be appropriate in tracing the presence of inequity in the delivery and distribution of resources, the rationality of which, at the very inception, was grounded within statistical analyses. If any other method, conceptually different from the statistical one, were applied for the analysis, the method itself might have influenced the outcome of the result. But now, not the method, but the ground realities will be responsible for the finding of empirical work.

There were certain dilemmas in how to narrate this work: Should I begin with theoretical generalizations leading to the empirical research and then to the conclusion, or, should I use "counter-generalization" by grounding the argument on empirical and factual observations supported with theoretical interjections? These questions do pose uneasy negotiations between the two at various stages of theoretical referencing. Otherwise, I have tried to keep the progression of the work rather straightforward with theoretical propositions made in Part I on positions and propositions, empirical research discussed in Part II on the delivery of urban water supply in Delhi, and concluding comments in Part III.

In Part I, the theoretical propositions are attempted at three levels. To begin with, a theoretical scaffolding is assembled by involving specific and significant conceptual strands on ideology, policy and governmentality, and the conception of justice and equity is understood in relation to this framework. A theoretical grounding is then made by relating the IPG framework to the context of the democratic India. The notion of *planning as a policy*, adopted in the independent India, is discussed. Clear references are made to the urban planning policies inspired by the normative ideals of modernist versions of the West. Furthermore, the MPD 1962 is analyzed by revealing traces of governmentality techniques within the rationalities of the Plan itself. At the end, to read conditions of injustice and inequity in distributions, a theoretical alliance is approached by bringing together fourfold arguments: the "strategic-relational" interpretation of the selective distribution by the state, the political-economy arguments of the location-specific and the growth-centric distributive decisions, and the poststructuralist politico-legal notion of the *state of exception*.

Part II, essentially containing the empirical research on the delivery of water supply in the context of Delhi, begins with discussing relationships between infrastructure, planning, water supply, and the city. Research methods and the case study selection process are, then, touched upon.

Quantitative analysis establishes that the property tax categories, devised by the MCD, can be considered as a surrogate indicator of the socioeconomic strata in Delhi. The empirical work essentially examines the notion of inequity and injustice in water supply by the DJB. A threefold argument, based on the critiques of the planned provisions and quantitative analyses, is used to establish relationships between water supply by the DJB and the socioeconomic spaces. At the end, patterns of the political funding on water, namely the MLALAD fund, are observed with a view to understand its bearing on socioeconomic spaces.

The concluding Part of the book reiterates certain notions about the equity in the distribution of basic services, the performance of urban management, and the city-reading possibilities as the alternative to the *two-city* theories of binary opposites.

Notes

1. Marxist, Marxian, and marxist are the three conventions to describe the theoretical reference to Marx's works. Marxist is the work that follows Marx's positions. Marxian is a rather comforting academic notion popularized by scholars like David Harvey, Neil Smith, and others to describe the theoretical connections with Marx's world-view, whereas marxist is applied in a particular way, for example in Neil Smith's work, of constructing a critique. I prefer to use the expression, Marxist, without any such fine categorical or methodological differentiation.
2. Various dimensions of similar apprehensions are discussed in detail by Ekers and Loftus (2008) in establishing "dialogues" between Gramscian and Foucauldian frameworks with relation to generic water discourses.
3. Comparisons of works of Althusser and Foucault are made elsewhere as well (Hall 1985, Goldstein 2004, Montag 1995).
4. Larner (2000) uses ideology, policy, and governmentality as three classifications to sort out interpretations of neoliberalism in various writings; however, he treats these taxonomies more as explanatory tools and not as the generic theoretical framework and hence, has theoretical references completely different from those in my work.
5. One must not mix up the way politics is used here with the term, popularly referred to the activities of the political parties in representative democracy. Politics, here, represents the *disagreement* with the policy (or the police).
6. In an article, entitled "Cities and Diversity" (2005), Susan Fainstein defines "the just city in terms of democracy, equity, diversity, growth, and

sustainability." In a recent book, *Just City* (2010), Fainstein has further drawn from her earlier works on the issue as well.

7. In this connection, Sen's (1999) reference to the relation between famine and democracy is interesting. Sen, while citing the incident of the Bengal Famine that happened in the period of July–December, 1943, with a recorded toll of over 1.3 million deaths, argues that "no major famine has ever occurred in a functioning democracy" (ibid.: 342). He observes close connections between famines and authoritarian rules; e.g., colonialism (in British India or Ireland), one-party states (Soviet Union in the 1930s, China and Cambodia), and military dictatorships (Ethiopia and Somalia) (ibid.: 338–345). Policies of the government in terms of its accountability to people as well as the freedom of media reporting and public criticism are what give political incentives to the government to do its best to eradicate famine (ibid.: 343–344).

8. Kaviraj (2005) coins this term of "Nehruvian State," which I use here rather uncritically. Kaviraj (ibid.: 13) also feels that the "slow and insubstantial economic growth" was a hindrance to any serious income redistribution in the Nehruvian State.

9. Later, Fainstein (2005[2004]: 121), too, identified "the objective of planning as conscious creation of the just city, which requires a substantive normative framework."

10. Selective delivery indicates the delivery of the policy. When the state chooses to selectively deliver a policy to one set of people and not to the other or to one space and not to the other, even when the policy is same for both the groups of people and the spaces, it leads to inequity. Such conditions of inequity and unevenness are the outcome of selective delivery decisions made by the state on its own. If the state is selective about a space in delivering its policy, the spatial selectivity of the state gets revealed and when it is about a group of people with certain income and social status, socioeconomic selectivity of the state comes to the fore.

Here, Roy's (2003, 2005, 2008, 2009) assumptions are quite relevant: The informal decision-making by the state leads to such inequitable conditions.

11. Chatterjee (1994, 2004), too, brings out the oppositions between the conceptions of planning and politics in the context of the post-independent India.

12. Political society, a notion of "subaltern politics" as Chatterjee (2004) describes through various case studies, is formed by "mostly marginal and underprivileged population groups" that "transgress the strict lines of legality in struggling to live and work" and in doing so, make a number of connections with more influential groups of politicians, government staffs, and others to be able to manipulate certain conditions of living in their favor by using their voting power as the *strategic* instrument of the democratic political practices in India (ibid.: 40).

PART 1

Positions and Propositions

2

Ideology, Policy, and Governmentality

Policy, while intending to achieve an equitable distribution, does not necessarily ensure it, and the delivery, in reality, is modified by the politics of distribution. Two conditions often occur: one, when different set of policies are adopted for different/same sets of people, and the other, when certain deviations from the policy happen in the course of implementation.[1] Ground conditions may also cause certain manipulations of the policy implementation favoring particular space/s or group/s of people. I wonder whether the policy regime, for example, urban planning, may also have the room inherent in it for such manipulations.

Any selectivity, indeed, reflects biases of the state and does not portray the ideology of any democratic set-up to attain equality and liberty. Also, a policy, when decided, intends to achieve an idealized vision and is unlikely to have the rationality of the selective and differential unfairness. Hence, there has to be either multiple, sometimes contradictory, rationalities embedded within the policy itself, or special ground conditions influencing the delivery decisions on case-to-case basis. Both are the techniques of governmentality.

My first proposition is:

Ideology: justice idealized; Policy: justice rationalized; Governmentality: justice realized.

I discuss significant theoretical concepts involving ideology, policy, and governmentality, establish broad interrelationships among those, and finally, explain equity and justice in distributional possibilities (see Figure 1.2).

Ideology, Ideological Apparatuses, and Democracy

The purpose, here, is not to make any attempt to define ideology. Instead, I look at notions of ideology, rooted in Marxist positions, which are relevant and central to my arguments. Louis Althusser's readings of Marx's *Capital* (1970) and his essays (1994[1970]) are the key references here for absorbing the notion of ideology.[2] In Althusserian observations, ideology is what science is not; therefore, ideology is a binary opposite of science. His central thesis, "Ideology interpellates Individuals as Subjects," is arrived at by discussing two assumptions he puts across on ideology: one, "Ideology represents the imaginary relationship of individuals to their real conditions of existence" and the other, "Ideology has a material existence," which I will touch upon here (Althusser 1994[1970]: 123–126).

[I]imaginary relationships of the individuals to their real conditions" are what Althusser relates to the worldviews or the outlooks of life, for example, political ideology, legal ideology, etc.; hence, ideology is not the systems of "real relations" (ibid.: 123).[3]

It is representations or ideas of realities that may not essentially exist on ground in totality. Ideology constructed by the representations or the ideas, Althusser argues, has material exi:tence in "an apparatus, and its practice, or practices" and has no ideal or spiritual existence (ibid.: 126). While discussing material existences of ideology, he introduces the notion of ISA (Ideological State Apparatus). One may take an example of the notion of an institution like a school to explain the concept of the ISA. Ideology of the school (the apparatus), for example, exists in its *personnel* or the human resources like students, teachers and staffs; in its *equipment* of building, books, and other necessary articles; in the *organization* of relationships, duties, and responsibilities; and finally in its *rituals* of classes, lectures, annual festivals, sports events, etc. Personnel and equipment have more tangible material existence and rituals are the practice, whereas organization can be seen both as material and its practices.

Alongside the ISA, such as the religious ISA, educational ISA, family ISA, legal ISA, political ISA (the political system, including different parties), cultural ISA, etc., Althusser also presents a completely different state apparatus: the "Repressive State Apparatus," (RSA) such as the government, administration, army, police, prison, etc. Primary operations of the RSA are through repression and secondary ones through ideology, whereas for the ISA, it is the reverse; the major functioning is through ideology and the minor functioning is though repression. These subtle distinctions between the two apparatuses may often help the state, which, depending on its ideology at that point in time, may interchange the role of these apparatuses in practice (Althusser 1994). If the ideology of the state, as understood by Althusser, is the ideology of the ruling class working primarily under the ideology of the state, then ISAs are highly exposed to the activities of the ruling class (ibid.). The practice of these apparatuses is what I recognize as the *policy of the state* and through it, the manipulation and the control of the state happens. Theoretical possibilities of connecting Althusserian concepts of ideology with Foucault's notions of governmentality open up.

Althusser's propositions of ideology received many criticisms as well (Abercrombie et al. 1994[1983], Bordieu and Eagleton 1994[1991], Hall 1985, Pêcheux 1994[1982]).[4] One significant criticism has come from the prominent literary critic Terry Eagleton (1994[1991]: 217), who feels that Althusser's model of ideology is "too monistic". Yet, he recognizes the pioneering theoretical breakthrough that the Athusserian model has managed to make. However, it is widely accepted and contested that the interlocution with Marx in Althusser's writings in founding the theory of ideology is the starting point of many iterations on similar concepts (Balibar 1992, Hobsbawm 2011, Zizek 1994).

Let me introduce the notion of democracy as the setting to take this discussion ahead. Democracy is both ideological and political; it is ideological because of its foundations on the assumptions of equality and liberty (also, *égaliberte* for Balibar 1994), and it is political for giving space for "political participation, dialogue, and public interaction" as necessary practices and procedures of democracy (Rancière 1999, Sen 2009: 326). Equality is the very premise of democracy and the ontologically given condition of politics (Rancière 1999, Swyngedouw 2009).[5] The same applies to the conception of liberty as well. The unrestricted, rather ideological, idea for justice is the convergence of equality and liberty ensuring the absence of unfairness or inequity and repression, respectively. This I understand and refer to as the *just condition* here.

Democracy is also seen as a formal "exercise of public reason" (Habermas 1995, Rawls 1999, Sen 2009). John Rawls, a leading figure in moral and political philosophy (1999), mentions that "[t]he definitive idea for deliberative democracy is the idea of deliberation itself. When citizen deliberate, they exchange views and debate their supporting reasons concerning public political questions" (quoted in Sen 2009: 324).

In such a setting, where opportunities and choices seem equal, at least conceptually, "social justice and a better and fairer politics" can be expected (Sen 2009: 351). Politics, which Althusser understands as an ISA, functions in pursuance of the ideology of democracy, more precisely, the notion of equality and liberty leading to the condition of justice. Practice of such an ideology by the state guides the choice of its (state's) policy, which tends to differ from the policy of the state under other ideological conditions. Erik Swyngedouw (2009: 605), an eminent planner–geographer, refers in his discussions on the "postpolitical city" to the visions of Etienne Balibar and Jacques Rancière in identifying politics as articulations of the right to equality, and also as demands of exposing the "wrongs of the police order." It directs me to the discussion on policy.

Politics of Democracy vis-à-vis Policy

"What makes politics an object of scandal is that it is that activity which has the rationality of disagreement as its very own rationality," Jacques Rancière argues in his book *Disagreement* (1999). Subsequently, he identifies how "democracy differs from practices and legitimizations of the consensus systems" (ibid.: xii and xiii). Democracy, to him, is "the institution of politics itself" and "every politics is democratic" which writes, argues and, rewrites in a cyclic process of actions on the premise of equality in which the very notion of equality is always tested. Therefore, democracy is not about consensus:

> Democracy is not a regime or a social way of life. It is the institution of politics itself, the system of forms of subjectification through which any order of distribution of bodies into functions corresponding to their "nature" and places corresponding to their functions is undermined, thrown back on its contingency. (ibid.: 101)

Iris Marion Young (2011[1990]), too, puts politics at the core of the discourse of justice in her influential book, *Justice and the Politics of Difference*. For her, politics broadly stands for institutional arrangements;

> ...the concept of justice coincides with the concept of the political. Politics as I defined it ... includes all aspects of institutional organization, public action, social practices and habits, and cultural meanings in so far as they are potentially subject to collective evaluation and decision making.

Politics in this inclusive sense certainly concerns the policies and actions of government and the state, but in principle can also concern rules, practices, and actions in any other institutional context. (Young 2011[1990]: 34)

Her concerns are, however, more generic. I would, at this point, continue with Rancière's (1999) discussions for identifying the difference between politics and policy, conceptually.

Rancière (1999: 101–102) introduces a term called "post-democracy" to indicate the practice of "consensus", which is inconsistent with the very concept of democracy of contradictions and oppositions of ideas and opinions. He relates this consensual display of the "conceptual legitimization of a democracy" with the government practice, which reduces the democratic operations to the "police management" of relationships between the mechanisms of the state and the combinations of social interests; and according to him, this is the "logic of police order" (ibid.: 102–108). The consensual policy arrangement, thus, converts the "political" to "policing", to "policy-making", to managerial consensual governing (ibid.). Such a notion clearly separates "politics" (*la politique*) from "the police" (*le police*) (Rancière 1999, Swyngedouw 2009).

> Politics is generally seen as the set of procedures whereby the aggregation and consent of collectivities is achieved, the organization of powers, the distribution of places and roles, and the systems for legitimizing this distribution. I propose to give this system of distribution and legitimization another name. I propose to call it the *police*. (Rancière 1999: 28)

Therefore, democracy, the ideology, is seen standing apart from the democratic policy, the police.

The perception of police is often related to a proper order of governance (Dikeç 2005, Swyngedouw 2009, Urbinati 2003). The police is "an order of bodies that defines the allocation of ways of doing, ways of being, and ways of saying, and sees that those bodies are assigned by name to a particular place and task" (Rancière 1999: 29). This notion of police is quite close to the rationalities of distribution, for example, planning.

The police, not being a social function, refers to the activities of the state and the ordering of social relations, both of which are usually related to the policy of the state. In such a sense, policy of the state can be viewed as the practice of ideology of the state in which both the Althusserian state apparatuses come together. Policy, meaning the plan of action or ways of government and administration often stated by the state, is in continuation with the discourses of the police. The word, "policy," originated from the French word *policie* (civil administration), interestingly shares similar etymological connections with the word, "police." Policy is also *prudent* or thoughtful *conduct* and, therefore, consensually arrived at after many deliberations by the state depending on its ideology.[6] For example, in the post-independence Indian democracy, issues of equity were addressed through policies of comprehensive planning for which the Planning Commission was established. If inequity is the problem, planning is the solution. This is a condition for "post-democracy" that identifies the problem, generates the opinion, and achieves the consensus on the most logical solution without involving the contradictions of the politics of democracy (Rancière 1999: 107). These policies, then, help the state to set mechanisms to control and govern its people and space leading to the governance.

Governmentality and Techniques of Delivery

Governance is often mentioned in association with mechanisms and/or systems of organization for the coordination of activities to solve certain problems and, thus, is involved with the delivery ends of the binding policy decisions (Swyngedouw 2009, Urbinati 2003). The recipients are not "the people" as a collective political subject or "the citizen" as a theoretical notion of the state, but "the population", an empirical category of unidentifiable masses derived by the techniques of the policy, such as census or sample surveys (Chatterjee 2004: 34, Rancière 1999: 80, 104, Swyngedouw 2009: 608, Urbinati 2003: 80). The very notion of applying necessary techniques of arrangements to provide for the well-being of the population leads me to the next discussion, on governmentality.

Here, I attempt to expand upon ideas embedded in Michel Foucault's discourses on governmentality (Barnett et al. 2008, Burchell et al. 1991, Chatterjee 2004, Corbridge et al. 2005, Dean 1999, Donzelot and Gordon 2008, Foucault 1975[1972], 1984, Larner 2000, Legg 2005, 2007, Lemke 2001, 2002, 2007, Maidan 2008, Parrott 2002 [1996]; Rose, 1999). Governmentality appears to combine three aspects: the power, "the tendency" of the emergence of government as a type of power leading to certain apparatuses and knowledge; the analytics, "the ensemble" of exercising very specific and complex kind of power, shaped by the institutions, procedures, analyses, and tactics; and the governmentalization, "the process" of transformation toward the administrative state (Foucault 1978 in: Burchell et al. 1991, pp. 102−103).[7]

Government is referred to as the "conduct of conducts," of the self and others in leading, directing, guiding, or calculating the methods behind all of these (Dean 1999: 17, Lemke 2002). The government is concerned with men and their relationships with things such as "wealth, resources, means of subsistence, the territory with specific qualities, climate, irrigation, fertility, etc.," with their way of life, customs and habits, and with situations like calamities and misfortune of famine, flood, death, etc. (Foucault 1978: 8, 10, quoted in Foucault 1984: 16). These concerns of the government lead to complex activities, methods of its operations, or techniques which is called governmentality. Governmentality is a kind of rational and calculated activity that is carried out by multiple agencies that engages diverse techniques and knowledge and tries to "shape conduct by working through the desires, aspirations, interests and beliefs of various actors ... [with] unpredictable consequences, effects and outcomes" (Dean 1999: 17–23).

"The art of government," Foucault reminds us, is concerned with the "management of the state" (Foucault 1984: 15). Governmentality, then, can be seen as the rationalities or mentalities of the government and the practices influenced by the same (Dean 1999, Lemke 2002). For example, if the distributional practice of a scarce resource, like water supply, chooses to oblige higher strata of society more, it would expose certain rationalities of such decisions and mentalities behind those. The mentalities behind practices of the government encompass a collective activity of thinking based on accepted bodies of knowledge. Such knowledge may not stem from any collective consciousness of higher level, but from a common sense, and is "taken for granted" by its practitioners (Dean 1999: 25). What is of significance here is the resemblance between the collectively agreed upon thinking activity of governmentality and the consensual post-democracy in Rancière's notion of the police. Governmentality, then, extends into the rationality forming the policy. At the same time, this rationality, arrived at through processes of consensus, is practiced without often being questioned and, therefore, tends to be normative. Normative tendencies, thus, have a significant impact on the practice at the delivery ends of the governmentality.

The mentalities of the government to control the self and others include *rational* aspects from fields such as psychology, economics, management, or medicine, as well as *a-rational* ones from politics and vocabulary, both combined into a rational technique of governmentality. To explain this point, Dean (1999: 25–26) uses the example of strategies of self-control on health, including the dieting regimen based on the scientific knowledge and the meditation or the food habits coming from the spiritual and cultural understanding. At the same time, the dieting regimen is a technique to control others' heath as well. This aspect of governmentality also indicates *the tactics of fire-fighting* in it when the logic of governing comes from variable conditions of ground realities and, consequently, policies are worked out.

Here, I do not intend to delve into the binary comparisons of oppositions and similarities of larger premises of Foucault's notions with Marxist theoretical constructs as discussed elsewhere (Balibar 1992, Donzelot and Gordon 2008, Ferguson and Gupta 2002, Hall 1985, Harvey 1996, Lemke 2001, 2002, 2007, Olssen 2004, Zizek 1994). Instead, I see a point of convergence here between Althusser's notions of ISA and Foucault's disciplinary techniques in governmentality (Balibar 1992, Lemke 2001, Montag 1995). Foucault's construct of power forming itself "from below" is often seen opposed to Althusser's supposition of the state as the large ideological edifice (Zizek 1994). Interestingly, both coincide with the processes of the policy formation (Ranciere's conception of "police"), albeit from different ends and notions.

Processes of the formation of policy, for example, equitable distribution, is approached from two ends: One, from the ideological end, in the creation of the ideal condition, that gets compromised through the consensual rationality in the formation of the policy; and the other, from the governmentality end, in the negotiation with real conditions through the formulation of rationalities as accepted policies to be practiced (see Figure 1.1).

Social Justice and Possibilities of Distribution

So far, I have elaborated upon the notion of democracy as an *ideological formation of equality and liberty*, which is open to active and vibrant *politics of disagreement*. Democracy may also be accepted as a necessary condition of justice.[8] While the ideals of equality and liberty are the foundation of democracy, the content in democracy provides a space for the formal "exercise of public reason" (Balibar 1994, Chatterjee 2004, Fainstein 2006, Habermas 1995, Rancière 1999, Rawls 1971, Sen 1979, 2009). As the meeting point of "political participation, dialogue, and public interaction," democracy offers an idealized setting of equal opportunities and choices and enhances social justice, perhaps, through "a better and fairer politics" (Rancière 1999, Sen 2009: 326, 351).

Practice of such an ideology by the state leads to the choice of its policy, which tends to differ from the policy of the state under other ideological conditions. For example, one-party rule, military dictatorship or monarchy, or similar other authoritarian rules have a different ideology than democracy and they formulate policies accordingly.

Democracy, being an ideological condition of justice, gives a broad context within which I can now reflect upon the concepts of social justice and accordingly, distributional possibilities in particular. My major references are Rawls' (1971) and Sen's (2009) works on the idea of justice. Rawls (1971) proposes on the line of the traditions of the "social contract," the pioneering concept of "justice as fairness" with equality and liberty as the foundation of the just society. As against Rawls' suggestion of the rational choice in deciding the principles of social justice, Sen (2009) brings out the issue of "actual choice" when one is to choose from more than one rational alternative and introduces the idea of "the capability approach" (ibid.: 183). While arguing for the capability approach in understanding the idea of justice, Sen too, acknowledges that the idea of fairness is "central to justice" (ibid.: 62).

The assumption of the course of my discussion here is: The notion of "justice as fairness" is the policy framework based on consensus, and the methods of the "capability approach" is the technique of governmentality to assess the actual delivery at the end.

"Justice is the first virtue of social institutions, as truth is of systems of thought. (. . .) [I]n a just society the liberties of equal citizenship are taken as settled" (Rawls 1971, in Stein 1995: 64). Rawls' (1971) pioneering theoretical formulations on social justice have been the most significant framework, to be followed by several other important works.[9] Based on the established traditions of "social contract," Rawls' proposition, "justice as fairness," highlights two elements: "(1) an interpretation of the initial situation and of problem of choice posed there, and (2) a set of principles which, it is argued, would be agreed to" (Rawls 1971 in Stein 1995: 69). What is just and unjust is a notion in dispute, and people disagree on the principles of determining various conceptions of justice within a similar democratic setting. Foucault's interpretations on possibilities of justice or justices underline this concern as well. Politics of democracy in Rancière's perspective, too, talks about the disagreement on the perceptions of equality and liberty (or *égaliberte* discussed in Balibar 1994).

"Justice as Fairness" appears to be a policy based on the *rational choice* in determining principles and arrangements for social justice. Principles of social justice, in turn, "provide a way of assigning rights and duties in the basic institutions of society and ... define appropriate distribution of the benefits and burdens of social cooperation" (Rawls 1971 in Stein 1995: 64; Figure 2.1). "The political constitution and the principal economic and social arrangements" are the major institutions he refers to (ibid.: 66). The "concept of justice," to Rawls, is *fairness*, which he distinguishes from the "possible conceptions of justice" in the same setting (ibid.: 65). The central idea of *fairness* is "a demand to avoid bias in our evaluations" and "for impartiality" with "unanimous agreement" (Sen 2009: 54, 59). The notion of *fairness*, thus, becomes somewhat similar to that of the *consensus agreement* in Rancière's works or the *common sense norms* in Foucault's. Rawlsian notion of justice is rationalized toward policy-making practices: if inequality is the problem because of unfairness, fairness is the just solution. Rawls (1971) recognizes it by positioning his notion of social justice at a lower level than what he calls "social ideal."[10] This is Rawls' reference to ideology which, I understand, tries to operate through the ISAs for an *idealized* notion of justice.

If ideology is justice idealized, policy is justice rationalized. Hence, Rawlsian notion of social justice can be comprehended as a policy.

Although the rational choice is important for analyzing the *actual choice*, what happens when one is to actually choose from more than one rational alternative? Sen (2009) questions the premise of the rational

Figure 2.1:
Concept of Justice as Policy: Rawlsian Formulations

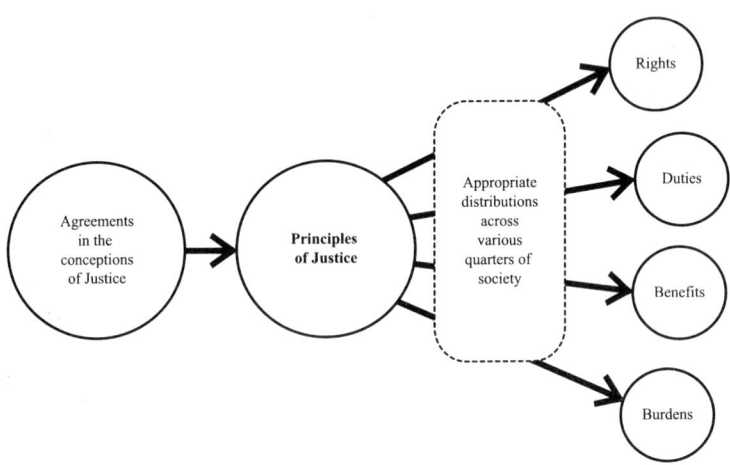

Source: Author.

choice in deciding the principles of social justice that Rawls suggests. In continuation with his original ideas of justice as fairness, Rawls (1993: 291; 1971) reiterates the principles of justice as:

a. Each person has an equal right to a fully adequate scheme of equal *basic liberties* which is compatible with a similar scheme of liberties for all.

b. Social and economic inequalities are to satisfy two conditions. First, they must be attached to offices and positions open to all under conditions of *fair equality of opportunity*; and second, they must be to the *greatest benefit for the least advantaged members of society*.

(Quoted in Sen 2009: 59; emphasis included)

The first principle emphasizes the uncompromising essential condition of personal liberty. The second principle has two components: one is concerned with institutional arrangements of ensuring opportunities accessible to all, and the other, also known as the "Difference Principle," is with the "distributive equity" and the "overall efficiency" toward benefitting the worst-off members of society (ibid.: 60).

In such a scenario, what is the *just distribution* of urban services and how is the access to such services ensured across diverse socioeconomic groups and spaces? According to the eminent urban geographer David Harvey, the concept of "a just distribution justly arrived at," indeed, outlines the notion of "social justice" (1975[1973]: 98). In his pioneering book, *Social Justice and the City*, Harvey (1975[1973]) discusses eight criteria of the just distribution based on "what is it that justifies individual making claims upon the product of the society in which they live, work and have their being" and finally suggests three criteria in the order of importance toward attaining social justice: *need, contribution to common good*, and *merit* (ibid.: 100; Rawls 1969, 1971, Runciman 1966). Harvey then brings out the sense of "territorial social justice" with respect to the historical–geographical materialism:

1. The distribution of income should be such that (a) the needs of the population within each territory are met, (b) resources are so allocated to maximize inter-territorial multiplier effects, and (c) extra resources are allocated to help overcome special difficulties stemming from the physical and social environment.

2. The mechanisms (institutional, organizational, political and economic) should be such that the prospects of the least advantaged territory are as great as they can possibly be.

(Harvey 1975[1973]: 116–117)

These points require more introspection and theoretical explorations, especially the first one on the issues of efficiency and distribution. The second point is the reiteration of Rawls' second principle of justice in spatial terms. However, both observations raise pertinent issues and factors related to social justice, namely "the pattern of territorial political power," the "allocative mechanism" of institutional and organizational entities, governing rules for "the pattern of territorial negotiations," and the relevance of "social and physical environment" (ibid.). Arguments on similar theoretical route are carried forward in the field of planning as well (Bromberg et al. 2007, Deakin 1999, Krumholz and Forester 1990).[11] Urban planner and economist Isabelle Fauconnier, too, considers the following while discussing about water supply and sanitation issues:

Equity may denote ideas of social justice, equality, and fairness across groups. Because it [equity] is an inherently subjective ideal—and not an ideal for everyone—it has often remained outside of the scope of economics, which itself remains dominant in the realm of public policy making. (Fauconnier 1999: 39)

The Rawlsian theory of justice continues to offer a broader conception of equity.

Iris Marion Young, a political theorist, has different concepts than Rawls', which requires a mention here. In her book, *Justice and the Politics of Difference*, she argues that justice is not confined to distribution alone, but encompasses "institutional conditions" leading to "two forms of disabling constraints, oppression and domination" (Young 2011[1990]: 39). As mentioned earlier, she considers that politics is what can capture the narrative of these institutional conditions and "[t]he scope of justice," she suggests, "is much wider than distribution, and covers everything political in this sense (ibid.: 34)." Young doubts that discussions limited to distribution pattern may not necessarily encapsulate matters related to "decision-making procedures, division of labor, and culture" (ibid.). She thinks that two key problems with distributive paradigm are: one, "it tends to ignore, at the same time, it often presupposes, the institutional context that determines material distributions," and the other, "when extended to non-material goods and resources, the logic of distribution misrepresents them" (Young 2011[1990]: 18).

For her, exploitation, marginalization, powerlessness, cultural imperialism, and violence are "five faces of oppression," where "oppression is a condition of groups" (ibid.: 39–65). She brings in "class" or "group" as the basic entity that encounters forms of oppression and power. Oppression usually emerges out of power relations—"who benefits from whom and who is dispensable" (ibid.: 58). "Social justice," Young suggests, "requires not the melting away of differences, but institutions that promote reproduction of and respect for group differences" (ibid.: 47).

Young's theorization goes beyond the specificity of distributive justice and becomes overarching. However, where class separation and corresponding injustice/inequity are distributed across spaces, equity and justice discourses conjoin with physical spaces, and distributive justice becomes relevant. Interestingly, the reliance of institutional practices, politics of differences, and power relations in understanding influential groups in her thoughts underlines its importance in my work.

Now, I shall touch upon Amartya Sen's theoretical formulations, discussed in his recent book *The Idea of Justice* (2009). Sen (2009: 231) introduces the idea of "the capability approach", developed by Philosopher Martha Nussbaum (1993), in which individual advantage is assessed by "a person's capability to do things he or she has reason to value." The concept of capability also gives necessary opportunities to a person who is free to make the choice of his/her liking. Sen claims that the main contribution of the approach is the due importance given to the information that influences "the assessment of society and social institutions" and such "informational focus" works for the advantage of an individual as well (Sen 2009: 233). The capability approach is conceived to be "a serious departure from concentrating on *means* of living to *actual opportunities* of living" (ibid. emphasis in original). This particular assumption underlines the difference of focus in Rawls' and Sen's approaches. For Rawls, equitable or just distributions of primary goods help in achieving quality of freedom and equality, whereas Sen argues that "the *means* of satisfactory human living are not themselves the *ends* of good living" and the focus should be on "the opportunity to fulfill ends and substantive freedom to achieve those reasoned ends" (ibid.: 234).

Interestingly, similar to the concerns of Rawls, Sen observes that it is the responsibility and obligation of societies to provide a person with opportunities of facilities and services (e.g., health care, water, etc.) without which one does not have the capability to enhance their conditions and, therefore, can be considered deprived (Sen 2009: 238). This convergence is what leads me to assume the following:

> *The pre-requisite for offering opportunities to enhance capability is to provide an access to means in terms of services and facilities, and if this is not addressed in the public provision by the state, it is inequitable leading to deprivation.*

For example, if water is not supplied to a group of people, they do not have opportunity to access basic services. It would be even more unjust if the poorer sections of society have to face conditions of no supply or inadequate supply of water from the state, as they would have limited capability to overcome the difficulty of supplementing the deficiency in supply from other alternative sources and means. The capability approach focuses on the ends and is suitable for comprehending the just delivery of public services (ibid.). Also, there would be an unfair opportunity when the facilities or services, like water supply, are provided in one place and not in the other, even if the institutional and the policy arrangement might have agreed to make the same (facilities or services) available equally for both the places—thus, Rawlsian argument of justice as fairness can be used as a situational comparison of justice.

Notes

1. Policies, when put into operation in specific spaces or for specific groups of people, indicate that the state is selective about the spatial and/or socioeconomic aspects in delivering its own policy. Here, it is assumed that any technical limitation, wherever found, is to be addressed to ensure the *just* distribution. Otherwise, it exposes selectivity in the choice of technology. Such technological selectivity is gradually becoming highly significant in accessing information and facilities.
2. Louis Althusser observes:

   > It is well known that the expression "ideology" was invented by Cabanis, Destutt de Tracy and their friends, who assigned to it as an object the (genetic) theory of ideas. When Marx took up the term fifty years later,

he gave it a quite different meaning, even in his Early Works. Here, ideology is the system of the ideas and representations which dominate the mind of a man or a social group. The ideologico-political struggle conducted by Marx as early as his articles in the *Rheinische Zeitung* inevitably and quickly brought him face to face with this reality, and forced him to take his earliest intuitions further. (1994[1970]: 120)

3. At another place, Althusser mentions in the same article:

[ideology] is not their real conditions of existence, their real world, that "men" "represent to themselves" in ideology, but above all it is their relation to those *conditions of existence* which is represented to them there. It is this relation which is at the centre of every ideological, i.e. imaginary, representation of the real world. It is this relation that contains the "cause" which has to explain the imaginary distortion of the ideological representation of the real world. Or rather, to leave aside the language of causality, it is necessary to advance the thesis that it is the imaginary nature of this relation which underlies all the imaginary distortion that we can observe (if we do not live in its truth) in all ideology. (1994[1970]: 124; emphasis included)

4. Some of the significant criticisms on Althusser's propositions of ideology are: From the notion on ideological conditions of reproduction (Pêcheux 1994[1982]), from the conception of ideology as beliefs (Abercrombie et al. 1994[1983]), from the opinion that Althusserian views are "too aristocratic" (Bordieu and Eagleton 1994[1991]), and from the view that Althusser's notion of the ideology of the rule of the state being the ideology of the ruling class is a "classic variant" of the Marxist theory of ideology (Hall 1985).

5. Rancière observes:

Politics, as we will see, is that activity which turns on equality as its principle. And the principle of equality is transformed by the distribution of community shares as defined by a quandary: when is there and when is there not equality in things between who and who else? What are these "things" and who are these whose? How does equality come to consist of equality and inequality? That is the quandary proper to politics by which politics becomes a quandary for philosophy, an object of philosophy. (1999: ix; emphasis included)

6. Rancière's use of the words, police or policing, has a neutral sense not loaded with any negative connotation as such, and, he supposes, it is not to be identified with the term "state apparatus" (Rancière 1999: 29). He criticizes "a certain confusion of politics and the police" in the notion of the state apparatus coming out of an assumed opposition between the state

and the society, the portrayal of the state as a machine, and the imposition of the rigid order of the state on the life of society (Rancière 1999: 29). Policing is not exactly "the 'disciplining' of bodies as a rule governing their appearing, a configuration of occupations and the properties of the spaces where these occupations are distributed" (ibid.:). Thus, by distinguishing the theoretical construct of the "police" from two similar theoretical constructs by Althusser on "State apparatus" and Foucault on "discipline," Rancière recognizes similarities between these three notions.

7. Michel Foucault (1978) in his lecture titled "Governmentality" defines governmentality as:

   a. The ensemble formed by the institutions, procedures, analyses and reflections, the calculations and tactics that allow the exercise of this very specific albeit complex form of power, which has as its target population, as its principal form of knowledge political economy, and as its essential technical means apparatuses of security.

   b. The tendency which, over a long period and throughout the West, has steadily led toward the pre-eminence over all other forms (sovereignty, discipline, etc.) of this type of power which may be termed government, resulting, on the one hand, in formation of a whole series of specific governmental apparatuses, and, on the other, in the development of a whole complex of savoirs.

   c. The process, or rather the result of the process, through which the state of justice of the Middle Ages, transformed into the administrative state during the fifteenth and sixteenth centuries, gradually becomes "governmentalized".

   (Burchell et al. 1991: 102–103)

8. This conceptualization has a subtle difference from Fainstein's (2005, 2010) definition of the "just city". I consider democracy is not (merely) one of the aspects of the just city, but the prerequisite for justice and equity in a modern society, and hence, for the just city. Young (2011[1990]), on the other hand, considers "democracy as a condition of social justice" (ibid.: 91–95).

9. John Rawls' proposition of "justice as fairness" has been the single most significant framework for the theory of justice. His notions of "social justice" have been the theoretical underpinning of several important works that followed thereafter including that of Harvey (1975[1973]). Recently, in his book *The Idea of Justice*, Amartya Sen critically evaluates Rawls' theory of justice. He takes clues from Martha Nussbaum's work, *The Quality of Life* (1993), in proposing "the capability approach" (Sen 2009: 231–238). Sen's discussions on the capability approach focus on "the opportunity to fulfill ends and the substantive freedom to achieve those reasoned ends" (ibid.: 234). Application of the capability approach has also been recognized in discourses on planning theory as a significant route for "establishing values appropriate to the just city" (Fainstein 2006).

10. Rawls (1971, in Stein 1995: 68) considers that "social ideal" is a higher level formulation: "a complete conception defining principles for all the virtues of the basic structure, together with their respective weights when they conflict is more than a conception of justice; it is a social ideal".

11. A well-known example for equity planning effort was the Cleveland planning department under the leadership of Norm Krumholtz.

3

Planning and Its Practices of Delivery

M y second proposition is:

Planning is not an ideology, but a policy that may be approached from ideology. While being implemented, planning tends to pick up normative techniques of governmentality.

The proposition is built around the notions and understanding of policy, planning, and urban planning. The democratic space of the independent India and Delhi, in particular, is the context of this discourse. The contextual reference is important as arguments are likely to address definite historical moments of ideology, policy, and governmentality while spanning across time.

Policy and Planning

Nehru in his great narrative, *The Discovery of India* (1994[1946]: 400), recognized the concept of free India moving toward "some of the fundamentals of the socialist structure," one of the essential ideals of which was to reduce the gap between the rich and the poor to a minimum. The overall vision of free India had ideological alignments with socialist aspirations and, within a democratic set-up of government, an objective of doing welfare for the society at large. At the outset, Nehru's vision (1994[1946]) in framing policies followed the idea of the universal welfare state, especially for the provision of goods and services by the state, which was conceived as a means of ensuring social well-being in the line of "Keynesian welfarism."[1] Ideology of the welfare state is often regarded essential for the well-being of the citizens (Veenhoven 2000). The reduction of inequality, the augmentation of the living standards, and the social integration are some of the vital objectives of the welfare state (Barr 1992). There is a view of two variants of the welfare state depending on the service provisions: the "universal welfare state," where services are to be provided for all socioeconomic groups, and the "residual welfare state" where public provisions, being "income-tested," are targeted for the benefit of the "poor" and services for the "non-poor" are mostly left to the private delivery (Willensky and Lebeaux 1965, in Barr 1992: 3). India very clearly began as the "universal" welfare state and has gradually got limited to the "residual" version.

Kaviraj (2005: 13–14) makes a couple of interesting points on this issue: The notion of the Nehruvian State, he considers, is a "poor people's version of the welfare state" and calls it a pure "statism" evoking "strong expectations of redistributions."[2] The expectation of redistributions had a convergence of two diametrically opposite, yet "real" aspirations for the relevant social groups: For the upper strata of society as "an instrument of economic growth, primarily for themselves and in the immediate future," and for the lower strata as "a promise of social dignity, an end of the caste system, and a distant dream of economic re-distribution" (ibid.: 14). Despite all criticisms, "the ideological commitment" of the Nehruvian State "to social reform and distributive justice" has always been its strength (ibid.: 13).[3] In such a state, the intended distribution should be what the state considers as just for the well-being of people at large.

Rigid state-control of the industries, public utilities, planning, and development process toward the modernization project were some of the key outcomes of the capital accumulation to be used for the betterment of the living standards of the society at large (Nehru 1994[1946]). The "equitable distribution" of resources (food, income, housing, health, education, utilities, etc.) was expected to ensure "national self-sufficiency" (ibid.: 396–398). And comprehensive planning, despite "having an air of unreality about it," was the single-most pervasive tool to achieve this goal (ibid.: 395). Nehru viewed planning almost like an ISA operating on the notion of equity to mitigate the problems of inequity. Planning was believed to have the ability to address the predicament of the production and the redistribution in the process of development (ibid.: 395–397).[4] A broad political "unanimity" was also observed, even among the "incongruous elements" in the National Planning Committee of 1938, in endorsing planning as the policy for the future (ibid.: 399). In continuation with the planning policies of the Nehru government, *Congress Resolution of the Avadhi Session* in 1955 also reiterated the role of planning at the core of the society;

> In order to realize the objective of Congress ... and to further the objectives stated in the Preamble and Directive Principles of State Policy of the Constitution of India, *planning should take place with a view to the establishment of a socialistic pattern of society*, where the principal means of production are under social ownership and control, production is progressively speeded up and *there is equitable distribution of national wealth.* (Gupta 1992: 199; WBPCC; emphasis included)

This alignment of the government and the political party was significant since it came from the platform of the Congress party, the largest political entity in India at that time. Approached from the ideological end of the policy-making processes, planning was underlined as the solution too, in bringing *consensus* among various national activities of economic, social, and cultural nature. Seen in light of earlier conceptual discussions on policy (or, police), planning was thus introduced in post-independence India as a policy.

For Nehru (1994[1946]: 400), socioeconomic conceptions of new India were "based on planning for the benefit of the common man, raising his standards greatly, giving him opportunities of growth, and releasing an enormous amount of latent talent and capacity," and he positioned it in the context of democratic freedom and cooperation. Sitting within a pre-independent India, he could visualize the methods of planning:

> Planning, though inevitably bringing about a great deal of *control and co-ordination and interfering in some measure with individual freedom,* would, as a matter of fact, in the context of India to-day, lead to a vast increase of freedom. We have very little freedom to lose. We have only to gain freedom. If we adhered to the democratic state structure and encouraged co-operative enterprises, many of the dangers of regimentation and concentration of power might be avoided. (ibid.; emphasis included)

Apparent manifestations of all these ideas could be identified with the setting up of the Planning Commission for the overall policy-framing and also with the mandate of planning cities for the new India. These observations indicated toward the delivery ends as well and can be seen in light of Amartya Sen's freedom-based capability approach, discussed earlier. Nehru recognized that the opportunities to enhance capability are the provision or access to means to be addressed in the public provision by the state for achieving equity by reducing deprivation. At the same time, he also highlighted the importance of empirical knowledge of the facts for forming the basis of proposed plans as well as the sense of discipline and control in the coordination among institutions (of the self) and people. In turn, methods of planning and their implementation tend to adopt some of the techniques of governmentality, and this forms the second part of my assumption. I shall elaborate this with specific references to urban planning.

Urban Planning Practices and Governmentality

I have discussed earlier the relevance of governmentality discourse in understanding the policy-making process with reference to norms of practice at the delivery end. At this point, one would like to apply certain key notions of the discourse on the techniques of governmentality to the context of urban planning, in general, and to India and Delhi, in particular.

Urban planning has contributed immensely in making the cities, especially since the beginning of the 20th century, under different ideologies of the state and through the masterly works of many famous architects, urban planners, and city designers.[5] Planning of/in the city and the urban context, both, are at the centre of my inquiries. In addressing the relationship of urban planning practices and governmentality, I start with discussions on the methods and effects of the practice of urban planning as the state policy, then, on the techniques of governmentality continuing from earlier discussions based on Michel Foucault's work and subsequent literature on that, and lastly, on the norms of practice at the delivery end of urban planning by taking the example of Delhi.

Practice of Urban Planning as the State Policy

Since the first half of the 20th century, urban planning has been a direct action of the state in bringing about sweeping changes in most of the cities we see today (Scott 1998, Taylor 1998). Predominant notions and values of urban planning in the late 1940s and 1950s somewhat fell short of inculcating nuanced perspectives of the then social, economic, or political realities (Alexander 1965, Colquohoun 1991, Jacobs 1964[1961], Krier 1990, Rowe and Koetter 2001[1983], Scott 1998, Stein 1995, Taylor 1998). Instead, the creation of the new and *improved* (sic.) way of life as a conscious departure from the existing patterns and systems of realities was the key aspiration of many countries and respective governments at that time. In search of new theories and practices representing the "Spirit of the Age" in art, architecture, literature, and other allied fields a set of values, technologies, and aesthetics emerged. It was the time that celebrated the pinnacle of the modern architecture, and urban planning was seen as its continuation. Most of the works and theories of town planning were by the master architects–planners.

The deterministic burden of establishing the new order for cities, perhaps, drove the master architect–urban planners, such as Le Corbusier, Tony Garnier, Jose Luis Sèrt, and others to decide finer details of the city design. Needless to say, most of these new city plans were conceived as the *clean slate design*, best depicted by Corbusier's scheme for *Plan Voisin* within the older areas in Paris. Simultaneously, along the lines of *Congrès International d'Architecture Moderne* (International Congress of Modern Architecture [CIAM's]) dictum of the "functionalist city" (included in the Athens Charter of 1933), emerged the strict mono-functional zoning of activities (namely live, shop, work, and play) in the city for which urban planners produced the "master plan" or the "land use plan" like a finished blueprint of a new city. Again, that was an end-state plan determining the land use and development guidelines till the plot level. These two approaches, I refer the first as *the utopia of grandness* and the second as *the utopia of detail order*, were extreme conceptual variations of the functionalist city with the utopia of creating new and better conditions of living and were applied comprehensively in the planning of two celebrated cities in the post-independence India, Chandigarh and Delhi, respectively.[6]

Influential leaders, like Nehru, and the technocrats, like Mahalanobis, both educated abroad, were also influenced by such mainstream worldwide trends in planning. Interestingly, Nehru's ideas of contemporary cities matched well with the ongoing paradigms of modern planning propagated in the West. The choice of Le Corbusier for the planning and design of Chandigarh and Albert Mayer and the Ford Foundation team for the First Master Plan for Delhi (MPD 1962) were indicative of the same. Planning of the post-independence Delhi was undertaken through statutory state mechanisms and resulted in the formation of institutions (like, the Delhi Development Authority) and instruments (like, the MPD) among others—a model to be followed later across the country. Instruments and techniques of planning of Delhi by the government are still being looked up to by several smaller towns as the most significant precedent.[7] However, it is pertinent to underline here that these two approaches of modern planning have remained as the most popular techniques adopted by the state in the 20th century. Both the planning practices of the utopian grandness and the detail order of the urban structure by the state and its products (the resultant cities) received many criticisms for its authoritarian determinism as well as for the lack of vibrancy and complexities in urban life (Alexander 1965, Colquhoun 1991, Jacobs 1964[1961], Krier 1990, Rowe and Koetter 2001[1983], Scott 1998).[8] To sum up, one may use phrases of architecture theoretician Alan Colquhoun that the city of utopian grandness is "artistic, ideological, and apodictic," whereas utopian city of detail order is "scientific, neutral and refutable" (1991[1989]: 107).

The utopia of modern planning has been connected with the redistribution and the reconfiguration of land by the state. After the state acquires land for "public purposes," previously existing complex land ownership patterns coalesce into a simplistic but gigantic blank piece of land, the *clean slate*, on which utopias of modern planning have been historically constructed and are still being repeated in many places through the state policy. Such practices of mapping, that simplifies the measurement and consolidation of land, produces centralized land maps, and sets up uniform property regimes for ownerships and taxation, are some of the *techniques of the state* to achieve legibility in governing its territory (Scott 1998).

This kind of practice of legibility connects with Foucault's notion of governmentality as well, where the land maps (whether the cadastral survey maps or the land use maps) become the tool of control and surveillance. At the outset, the planning policy of the post-independence Delhi is no exception to that. Also, most of the examples of both these variations of the utopian planning have initially been conceived as the apparatus of the ideology of the state (ISA), which, many feel, is to create better conditions for human living (Scott 1998, Taylor 1998). But, the urban planning by the state, eventually yielding rigid end-state applications, has got itself converted into the apparatus of repression (repressive state apparatus) of which ideology has reduced to an oblivious component.

Techniques of Discipline and Power in Determining Rationalities of Governmentality

Drawing on Foucauldian formulations, I shall now try to explain certain key techniques of discipline and power, such as totalization, individualization, and normalization, which are key notions of controlling (managing) the self and the other.

Since the 16th century, the problem of the controlling act, Foucault thinks, has been addressed by the state's power through a delicate combination of "individualization techniques" and "totalization procedures" (Rabinow in Foucault 1984: 14). Totalization procedures are addressed from the analytics of governmentality: the art of government and its statistics, the empirical knowledge of state's resources and conditions (ibid.: 16). Demography is a key component of such an information system. That is how men become the empirically quantifiable population used to form the basis of the policy-making. Individualization techniques then *classify/categorize* the population, *distribute* them over spaces, and often *dominate* the other through processes of "subjectification" (the way a human being treats himself or herself as a subject) (Foucault 1984: 11, Legg 2005).[9] The example of the "ladies compartment" of the local trains in India comes to my mind. Assigning of the "ladies compartment" presumes a concept of classification based on certain gender construct. It is also a distribution technique, and ladies, while using the designated compartment, turn themselves into the subject of the domination procedures. Techniques of both the individualization and the totalization process are vital components of Foucauldian formulations on discipline (Foucault 1975[1972]). Urban planning techniques, too, follow similar procedures, which one may see in the example of the MPD in a while.

Foucault's assumptions on governmentality explore *conditions of consensus* as well, which is evident in his notion of normalization (Lemke 2002). Normalization refers to "a system of finely gradated and measurable intervals" of distributions around *a norm* that controls, as well as, becomes the product of these "controlled distribution" (Rabinow in Foucault

1984: 20). Effects of a system of domination exercised by one group or element on another, Foucault thinks, runs through "the whole social body" (Foucault, quoted in Balibar 1992: 44). This idea of "the soul or the spirit" within the "social organism" is what Foucault calls "nominalism" (Balibar 1992: 44). Foucault's applications of the notion (and the metaphor) of the body in understanding the system, the organization, and the society at large is a unique reading coming from the understanding of "sexuality" as a complex and widely applicable phenomenon (Balibar 1992, Lemke 2001, Rabinow in Foucault 1984). This adherence to the "materialism of body" in Foucault's work, Balibar (1992) considers, differentiates his work from that of Marx.[10] However, this notion of nominalism is also what brings Foucault close to Marx's works, especially to the interlocution of Marx by the likes of Althusser (ibid.).[11] Later, Fontana and Bertani (2003: 277) reinforce this point, who feel that "Foucault maintained a sort of 'uninterrupted dialogue' with Marx, [who] was in fact not unaware of the question of power and its disciplines."

Foucault's panoptic view of the society in which conceptual evidences of the prison can be traced in other institutional edifices of society, such as schools, hospitals, factories, etc., is one such central idea in his works; also it leads to another similar one—the notion of disciplinary practices of control and surveillance of population (Foucault 1975[1972]). The apparatus of governmentality, as one understands, applies these disciplinary practices through intricate networks of the surveillance by which every bit of information on the life of population is monitored and controlled (Balibar 1992, Chatterjee 2004, Foucault 1975[1972]). Disciplinary practices, however, are not the law but the consensual norms of practice and this power of the norm is to be seen outside the power of the law (Foucault 1975[1972]). This particular arrangement of normalization would be useful in explaining the distribution practices of basic services across the so-called legal part of the city that I shall discuss later.

Foucault (1975[1972]: 190) identifies the technologies of policy formulation as "the measurement of overall phenomena, the description of groups, the characterization of collective facts, the calculations of gaps between individuals, and their distribution in a given 'population,'" and one can observe similar technologies in planning as well. These are exactly what Nehru also outlined in the methods of planning and till today, these are the techniques appropriated in undertaking any planning project including the ones that relate to the city, its spaces, and the people, and I shall elaborate that by taking the example of the planning of post-independence Delhi.

Post-independence Urban Planning in Delhi and Its Norms of Practice at the Delivery

After the independence of India in 1947, along with the adaptation of lots of existing administrative systems of the British Raj in the free India, the colonial capital city of New Delhi became the capital of the new democratic country. At this juncture, the city had witnessed a decennial growth of 90 percent of population when large number of refugees poured in due to the partition (Table 3.1).

Such a sudden increase of population led to the unplanned residential sprawl and growth of the informal sectors as its source of economy. To address these issues in the capital city, DDA was set up as per the Delhi Development Act 1957, to formulate the MPD with a view to "rationally" control the urban growth through comprehensive planned development. Post-independence physical growth of Delhi, as per the Master Plan has always been inside out by keeping the colonial capital city designed by Edwin Lutyens almost at the center of the concentric rings of more dense urban settlements. So far, DDA has prepared three key planning documents: MPD 1962, MPD Perspective 2001 (MPD 2001), and MPD 2021.

The MPD has been at the focus of many discourses: an exhaustive historiographical narrative of the MPD 1962 (Sundaram 2010), a historical account of the city of Delhi and related environmental concerns (Sharan 2006), evaluations and debates on the formulations of the MPD 2021 (Ribeiro 2007), and strategies taken by the authorities in making the Master Plan (Jain 2009). Several critiques also emphasize the shortcomings of the Master Plan and urban planning policies in terms of inadequate implementation of the Plan, for example, the insufficient provision for the housing for the poor leading to the resultant proliferation of slums and the status of the urban poor (Roy 2000, Verma 2002), the mechanisms of residential segregation and the factors at the micro-and macro-level (Dupont 2004), the formulation of inequitable policies in land, housing and transport (Kumar 2006, 2009), the lopsided transport strategies, land use

Table 3.1:
Evolution of the JJ Clusters–Squatter Settlements in Delhi (1951–1998)

Year	No. of JJ Clusters	No. of Squatter Household	Estimated Population (No. of Household X 5) (in lakhs)	Avg. Annual Growth Rate of Population (%)	10-yr Growth Rate of Population (%)	Population (in lakhs)	10-yr Growth Rate of Population (%)	Population of JJ Cluster / Total Urban Population (%)
			JJ Clusters^a			Delhi Urban Agglomeration^b		
1941						9.18		
1951	199	12,749	0.64			14.37	106.6	4.4
1956		22,415	1.12	75.8				
1961		42,815	2.14	91.0	235.8	23.59	64.2	9.1
1966		42,668	2.13	−0.3				
1971		62,594	3.13	46.7	46.2	36.47	54.6	8.6
1973	1,373	98,483	4.92	57.3				
1977		20,000	1.00	−79.7				
1981		98,709	4.94	393.5	57.7	57.29	57.1	8.6
1983		113,386	5.67	14.9				
1985	400	150,000	7.50	32.3				
1986		200,000	10.00	33.3				

(Contiued)

Table 3.1: (Continued)

Year	No. of JJ Clusters	No. of Squatter Household	JJ Clusters[a] Estimated Population (No. of Household X 5) (in lakhs)	Avg. Annual Growth Rate of Population (%)	10-yr Growth Rate of Population (%)	Delhi Urban Agglomeration[b] Population (in lakhs)	10-yr Growth Rate of Population (%)	Population of JJ Cluster / Total Urban Population (%)
1987		225,000	11.25	12.5				
1990	929	259,929	13.00	15.5				
1991			15.52[c]	19.4	214.4	84.19	46.9	18.4
1994	1,080	480,929	24.05	55.0				
1998	1,100	600,000	30.00	24.8		112.82		26.6
2001						137.83		

Source: Dupont (2008).

Notes: [a]S&JJ department, and food and civil supplies department, Municipal Corporation of Delhi; 1990 (January) and 1994 (March): based on direct surveys.

[b]Census of India, Registrar General of India (for population 1951, 1961, 1971, 1981, 1991, and 2001).

[c]Dupont's own estimation.

policies and related anti-poor investments (Tiwari 2003), environmentalist propagandas by the government and its agencies for "cleaning the city" and removing the slums (Baviskar 2002, Dupont 2008), and the analysis of legal verdicts and ongoing legal discourses on the politics of the removal of slums (Bhan 2009, Ghertner 2008).

Most of these discourses highlight the following: As a policy of the state, urban planning has its rationality of distribution, stated in successive Master Plan documents, which is faulty because of its limited understanding of complexities of the urban process, and, is often not implemented and delivered on ground to meet intended objectives. Here, I make an attempt to discuss the planning of post-independence Delhi in light of assumptions of governmentality mentioned earlier, and illustrate how some of the basic distributive strategies adopted in the MPD 1962 tend to demonstrate normative techniques of governmentality.

Core Concepts of the Master Plan

The draft MPD (ca. 1961) identified six key points, referred to as "Delhi Imperatives," to describe "concepts, measures and attitudes" of the Plan (ibid.: iv–v):

- Economy: The Plan had to achieve the desired results in the cheapest way possible and through certain conceptions, like; (1) self-contained districts (or, planning divisions) minimizing work–home relationships; (2) shifts in industries reducing space, equipment, capital installation costs for power, water and utility supply; (3) minimal interventions in the redevelopment of the "old city" and the re-densification of parts of "New Delhi Garden City," and (4) efficient architectural design and economical construction techniques.
- Architecture and civic design.
- Large-scale land acquisition: Acquired land would make the planning of the city possible and the land would be used as a resource for revenue generation to meet the development cost (the social good) thereby contributing to "the overall welfare of the city".
- Active social component: Planning was not to be limited to the physical improvement and, thus, planners were required to "deal with" people getting affected by the Plan, consider their problems, and communicate the Plan to them.

- Promptness in small projects for the citizen satisfaction in the Plan: Since with limited resources the considerable part of the Plan would take a long time, "simple and inexpensive" small actions were to be implemented to gain goodwill of the Plan.
- Determination and decisiveness in execution, coordination, negotiation, and completion of the Plan in order to bring good results.

Interpretations of the Core Concepts of the Plan

These six points indicated the policy of large-scale land acquisition to set out the process, and the rationality of spatial redistribution of income somewhat based on the locational relationship between home and workplace. Efficiency of the architectural design of spaces and the economy of the constructional technology were highlighted keeping in tune with the trends of the modern architecture being practiced in America and Europe and spreading itself as the "International Style" across the world at that time. The post-independence architecture in Delhi and in the rest of the country, too, would witness the proliferation of the International Style in building public and civic architectural edifices. The Master Plan document also hinted at possible building types and the logic of work space requirements for the multi-shift urban industries, which was, indeed, interesting. The first Master Plan, as opposed to popular perceptions, showed some bit of restraint by making minimal interventions in the "old city" as well as in the existing villages engulfed in the planned city. However, despite few attempts, redevelopment proposals were limited on account of inadequate framework to understand and handle such ground realities.

The most significant intent of the Plan seemed to be its social purpose: concerns for people, issues of slum redevelopment, notions of communication, and negotiation with people at large. It also recognized the need to involve social organizations, somewhat like the present day non-governmental organizations (NGOs), as an interlocutor between the government and the beneficiaries. The role of such organizations was to understand the social concerns of people, make them realize the benefits of a completely new notion of the plan, and get their feedbacks. The question of communication could have been a key point of the plan, but it required further explorations. Such an idea could have been extremely useful in assessing whether the public provision by the state to achieve equity was actually reducing deprivation. In implementation techniques, the plan preempted some smart moves in communicating its determination and decisiveness to the citizens through the idea of prompt execution of inexpensive projects. Some of these intentions of the Master Plan have often been overlooked by few of its critiques.

Somehow, these concepts and attitudes gave rise to a certain governmentality technique depicting the mentalities (and rationalities) of the establishment, that is, the planners and the government.

Techniques of Governmentality of the Master Plan

The Plan positioned the notion of *economy* (rather, economical) as the first imperative for the best execution. Foucault (1978) relates the most economical with the best and his concept of *government* establishes economy and order (of things and people) at all scale levels of the society. For him,

> the Art of government (. . .) is concerned with (. . .) how to introduce economy, that is the correct manner of managing individuals, goods and wealth within the family, (. . .) how to introduce this meticulous attention of the father toward his family, into the management of the state. (Foucault, 1978: 8, 10)

The MPD 1962 started along such dispositions of governmentality.

The Institution (the Panopticon)

> The Panopticon, was a design for a prison produced by Jeremy Bentham in the late eighteenth century which grouped cells around a central viewing tower. Although the prison was never actually built the idea was used as a model for numerous institutions including some prisons. Foucault uses this as a metaphor for the operation of power and surveillance in contemporary society. (O'Farrell 2005, 2007)

The utilitarian requirement of supervision, and the watch over synchronized networks of mechanism had needed, what Foucault (1979: 209) calls, a "panoptic arrangement," like the DDA. In general, DDA became the model of control over the *intended city* of Delhi. The power of such a model came through a number of disciplinary techniques of governmentality, like the surveillance and control, that formed the rationalities (or the mentalities) of the plan. DDA, thus, institutionalized the "ensemble" of governmentality through its technical knowledge of urban planning and had *a panoptic view as a central tower of observation* over the coordinated tasks required for planning the city.

Under the Ministry of Works, Housing, and Urban Development of the Government of India, DDA was formed as an 11-member body with the Administrator of the Union Territory of Delhi as the *ex-officio* Chairman (MPD 1962). DDA was empowered by Section (6) of the Delhi Development Act, 1957, to undertake the following:

> To promote, and secure the development of Delhi according to plan and for that purpose the Authority shall have the power to acquire, hold, manage and dispose of land and other property, to carry out building, engineering, mining and other operations to execute works in connection with supply of water and electricity, disposal of sewage and other services and amenities and generally to do anything necessary or expedient for purpose of such development and for purpose incidental there to. (Gambhir 1999)

The act also required the Authority as the centralized agency to carry out survey and prepare and implement the master plan. DDA acquired, held, managed, and disposed/leased the land for the purpose of planning the city. Delhi became the first urban area in India to have a comprehensive master plan in September 1962. Urban planning, being the public good and service, was conceived as the responsibility of the government authorities.

DDA had to coordinate, on the one hand, with the Ford Foundation team and the Town Planning Organization to prepare the plan and, on the other hand, with the civic authorities, such as the MCD and the then New Delhi Municipal Committee (NDMC).[12] The authority was to work with the central ministries for accommodating offices and staff housings and was expected to accommodate a huge number of refugees from the partition. Even today, DDA remains as the single most important organization for planning and urban development in Delhi, and is required to coordinate with multiple agencies (Figure 3.1). The overlapping of jurisdictions and complex urban governance in Delhi are also a matter of concern here.

Figure 3.1:
Example of Overlapping Governance in the Provision of Few Urban Services in Delhi

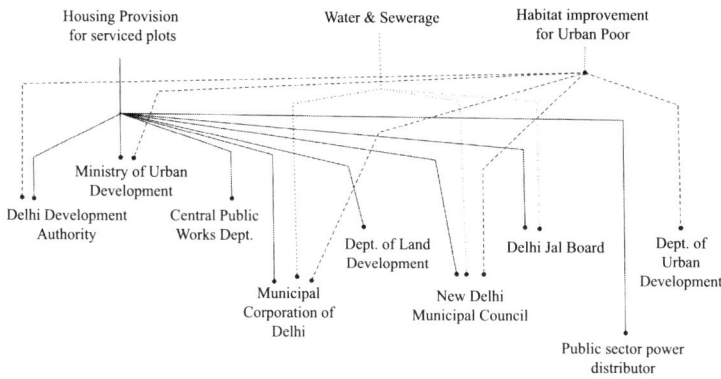

Source: Author.
Note: Singh and Shukla (2005), Jain (1990).

Land Control (the Sovereign and the Territory)

The sovereign signifies the undisputed authority of a king-like figure having the power to control its territory occupied by its subjects. Several theoretical and contextual interpretations of the *sovereign* and the *territory*, too, underline the control of the former over the latter (Legg 2005, 2007, Lemke 2002, Robinow in Foucault 1984, Scott 1998). Similar analogies are drawn about the state's authorization to DDA for the "sovereign control" of land, therefore the authority, to conduct the biggest land acquisition in India (Legg 2005, 2007, Sundaram 2010). In effect, a cumulative total of about 54,878 acres of land was notified by the DDA till October 24, 1961 (Gambhir 1999).

Somewhat different from the conception of the domination, another theoretical interface between the Marxist perspective and the Foucauldian notion recognizes the sovereign and the territory as mutually existent realities in the city. Sovereign controls the land along with the resources (including people) on it and gives a legal status to everything under its control. This position can be seen in the theoretical formulations of *citizenship* and *community* (Balibar 1991, Donzelot and Gordon 2008, Holston 2008), the creation and private appropriation of socioeconomic spaces (Lipietz 1980), the perspective of *legal* rights and *para-legal* social arrangements leading to the formation of the "civic society" and the "political society" (Chatterjee 2004), and the *formality* and *informality* in cities (Roy 2005). Several critiques of the Master Plan and prevailing urban processes of Delhi, too, have followed this conceptual route in constructing interpretations of the *two-city* theory of Delhi. I will revisit these discussions later.

In consonance with the notion of the post-independence India as a socialist republic, the land control remained with the state. The equitable and just distribution of the land and facilities on it would then be achievable, and possibilities of hugely varied land-holding by individuals would be nullified. This was the beginning of the shift from the pre-independent *estate* model of land ownership to the redistribution of land as the *public good*.[13] In the process, agricultural land was converted into urban land use. Also, the land acquired by the DDA was leased out to the private for residential or industrial uses, which would then generate the income and the surplus to be used for developmental work. Thus, the land was converted back to the *capital* again. A scheme called "Nazul Account II" (Chief Commissioner's Schemes) was opened to keep separate account of the receipts from land disposal through leasing etc., the expenditure on land acquisition and the surplus to be used for development (Gambhir 1999). The land was conceptualized as an ISA in which ideology of the public goods for the just and equitable distribution was the dominant factor and the re-conversion to the capital was a minor component as a RSA. The flattening of complex contours of ownership and characteristics of agricultural land by the state machinery, the DDA, into a single, large piece of clean slate, illustrates two sides of governmentality: first, techniques of control and surveillance as well as procedures of "totalization" and "normalization," both, with the help of conducting survey and producing maps, and second, the method of "sovereign" controlling the "territory."

The control of land taken over by the state, in this case the DDA, contributes to the accumulation of capital in two ways: first, by ascertaining the price of the land and the built property (for example, office, house or shop on that land), thereby getting the lease revenue mentioned earlier, and second, in an indirect manner, by setting up the regulatory mechanism of the land use recommendation to control the *proper use of land*. In effect, the regulatory mechanism has led to the complex and far-reaching legal and social consequences, captured in the theoretical formulations of *citizenship* and *community*, and in the variants of the *two-city* theory of Delhi.[14] This two-way land logic has given rise to the politico-legal accumulation process that very faintly started with Delhi within the welfare state and, within the "neoliberal market economy" of today, has reached the other extreme in Gurgaon at the outskirts of Delhi. The story of Gurgaon, however, is not the concern here at the moment.

City of the Master Plan (the Docile Body)

"A body is docile that may be subjected, used, transformed and improved" (Foucault 1975[1972]: 136). Urban planning of Delhi underwent similar stages. Through huge land acquisition, the city was subjected and then used, and the existing ways of living were transformed with a hope to achieve *a new and improved life.*

In the Marxist view, the accumulation of capital is mutually connected with the division of labor. For example, urbanization brings in a process of accumulation of men to one place, the city, which would then require the apparatus of production (educational, industrial, medical, etc.) to sustain them and vice versa. Interestingly, Foucault happens to make a direct reference to Marx on this issue. He introduces the notion of disciplinary techniques, which, while keeping surveillance on men and capital, makes "the cumulative multiplicity of men useful" and steps up the "accumulation of capital" (ibid.: 221). The "ensemble" of governmentality is created by the strong relationship between the apparatus of production, division of labor, and the disciplinary techniques (ibid.). In the MPD, the apparatus of production and the division of labor, both, were addressed through the method of land use recommendations and disciplinary techniques, through the production of the Master Plan and its rationalities.

Land use recommendations of the MPD allocated a particular plot of land to a use, such as, residential, commercial, industrial, and became *the apparatus of production* helping the urbanization process in accumulating capital and labor. At the same time, the plan created spaces to be used by the people for staying, working, studying, shopping, etc., and consequently, gave employment to the people as well. The plan became the mechanism that used disciplinary techniques, like hierarchical and spatial distributions of people and capital, in order to control the city. And the *intended city* of Delhi with this ensemble of governmentality was at work.

The understanding of "governmentality" discourse indicates that the government (the sovereign) having its control over the land decides what best is to be done with the resources and men on it (the territory). DDA used urban planning, which the state had thought the best option for the nation and its people, and produced the plan for the city of Delhi. The notion of "Docile Bodies," Foucault (1975[1972]) discusses in *Discipline and Punish,* may give an analogical reading of the city that the Master Plan projected, in which the city was conceived as the *body* subject to techniques of discipline and control. Discipline forms four kinds of characteristics in the bodies under its control:

> [I]t is cellular (by the play of spatial distribution), it is organic (by the coding of activities), it is genetic (by the accumulation of time), it is combinatory (by the composition of forces). And in doing so, it [discipline] operates four great techniques: it draws up tables; it prescribes movements; it imposes exercises; lastly, in order to obtain the combination of forces, it arranges "tactics". (ibid.: 167)

The city, the Master Plan had drawn up, was "cellular" having spatial distribution of eight self-sufficient planning divisions further divided into planning zones and residential neighborhoods accommodating activities and people. It was made "organic" by coding and controlling activities to be accommodated over spaces and "genetic" by including the locational decisions of labor and capital. Time taken for the movement of people and goods across spaces became important in the plan. And once the tactics of combining spaces, activities, and time were arranged together in the land use plan of Delhi, the city became "combinatory."

Land Use Plan (Combinatory Tactics)[15]

In governmentality discourses, the notion of population is conceived as "the empirical knowledge of the state" of its resources and people (Rabinow in Foucault 1984: 14) as well as the "undifferentiated atoms distributed through abstract space and time" (Curtis 2002) and in all practical purposes, population is "identifiable, classifiable and describable" as a group (Chatterjee 2004: 34, Legg 2005). Population is estimated by using "the science of the state"—the techniques of statistics with the help of censuses or surveys (Chatterjee 2004, Curtis 2002, Rabinow in Foucault 1984: 14). Population is used as the method of governmentality, on the one hand, as the "totalization" techniques converting individuals into targets of the policies of the state and, on the other hand, as "individualization" techniques classifying people (e.g., in terms of income groups) and distributing them over spaces (Chatterjee 2004, Curtis 2002, Foucault 1984, Legg 2005).

MPD 1962 used similar techniques of governmentality in population distribution based on income-driven categories, for example, higher income group (HIG), middle income group (MIG), lower income group (LIG), etc., and in the provision of the respective housing types. These population categories loosely underlined the *class* formation in the plan. However, at the time of real delivery, the housing provision fell short of its targets. The state has often been held responsible for the selective implementation of housing projects and, in turn, depriving the LIGs of the opportunity to live in the city (Kumar 2006, 2008, Verma 2000).

Land use plan was the single most important outcome of the MPD 1962. The rationality of such a plan was based on the "population projections" over the planned period, "decentralization" of the development through a poly-nodal model, and "distributions" of activities over spaces across scale-levels through hierarchy and zoning techniques. Population was the key empirical construct of the Plan. The strategy was to meet an anticipated gradual increase of urban population. The population density figures were used over the areas in the land use plan to indicate the distribution as well. At the outset, MPD's understanding of population was rather Malthusian, and perhaps, that was the reason why high population density in the older parts of the city was associated with the deterioration of the quality of living. Higher density, to such an empirical understanding, would mean more people on land, leading to smaller land-holding, less open areas, and more built-up spaces.

"In order to secure balanced development, and minimize frictions," "decentralization" of development was attempted through a poly-nodal model by creating eight planning divisions for urban Delhi, self-contained in terms of various land uses, and by proposing ring towns with sound industrial base and employment opportunities (MPD 1962: 7).[16] Within the regionalist model of the growth control of the city, the planning divisions became highly significant modules, territorial demarcation of which were initially based on the historical growth, character of development, intensity of land uses, and circulation pattern (ibid.: 64).[17]

A zone or division was divided into subzones. The idea of decentralization was to scale down the city in autonomous parts to establish "right relationships" between home and workplaces, shops and other facilities (MPD 1962: 7–9). Self-sufficient planning divisions created *individualization* of the atomistic modules of smaller city-parts, which were combined together to get back the large city. The decentralized model of Delhi also marked a strategic diversion from the socially stratified spatial model of "urban ecology." From the conception of "the city as living organism," this was the first level of spatial distribution resulting in "cellular" divisions (ibid.: 7).

"Hierarchical structure" in the MPD was conceived as the coding, control, and scalar distributions of activities (Figure 3.2; Table 3.2). Possible scale-levels of land uses of residential, commercial, institutional, leisure, and transportation were drawn up. Hierarchy was followed in the installation and distribution of both the physical (water supply, sewage, power, etc.) and social (educational, health, etc.) infrastructure. The pattern

Figure 3.2:
Space–Scale Relationships in the MPD

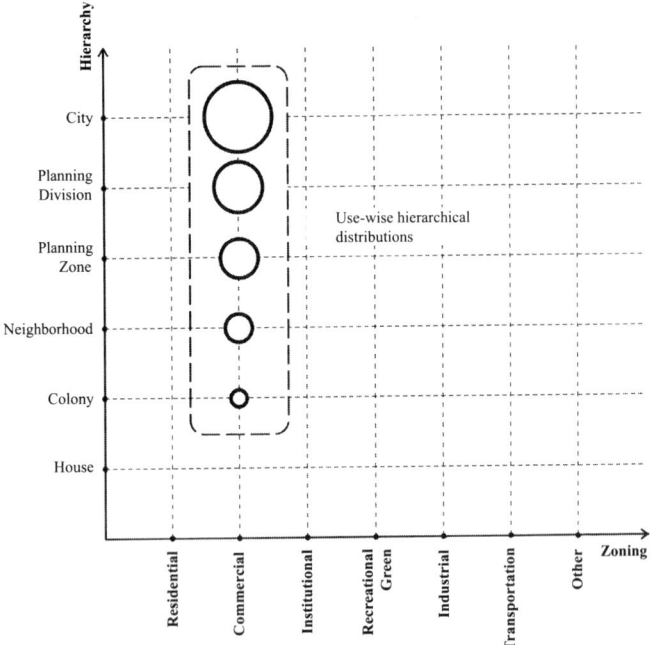

Source: Author.

Table 3.2:
Scale-levels of the Planned Community Structure: Master Plan for Delhi 1962

No.	Scale-levels	Population of Each Unit
1	Urban Delhi	–
2	Eight Planning Divisions (popularly referred to as zones)	3 lakh–7.5 lakh
3	Three Central Business Districts and 15 District Centers	Each District Center serving 1.5 lakh–2.5 lakh population
4	Community Centers	40,000–50,000
5	Residential Planning Areas	12,000–15,000
6	Residential Units	3,500–5,000
7	Housing Clusters	750–1,000

Source: Author.
Note: DDA (1962: 64–65).

of self-contained planning divisions was created from the "bottom upwards approach" with primary schools, high schools, community centers, and district centers forming the functional tiers, around which the community structure was expected to build up, with the lowest tier comprising the "housing cluster" with 750–1,000 people (MPD 1962: 64–65).

However, with the training, knowledge, and advice of the modern planning dogmas of the West, these clusters, in application, were planned around nursery schools and tot-lots. "Residential planning areas," in effect, became the derivatives of the "neighborhood planning" concepts. In the road network, too, definite hierarchy was established which finally gave rise to functional zones and plots as ordering elements of the city fabric.

Hierarchical planning with the distribution of live–work–leisure activities and related physical and social infrastructure facilities was the most important scalar notion of the MPD 1962 and, since then, it has persisted till date. Contrary to hierarchy, "zoning" in the MPD was conceived as the coding, control, and spatial (or horizontal) distributions of activities (Figure 3.2). The objectives for adopting the zoning technique were

> to promote public health, safety, and the general moral and social welfare of the community … to apply reasonable limitations on the use of land and buildings … to ensure that the city takes place in

accordance with the land use plan and its continued maintenance over the years. (MPD 1962: 44)

Having followed the dogmas of the "functionalist city" of the post-CIAM modernist physical planning, zoning became the central idea of organizing the city through the distribution of the predominant monofunctional land uses, such as residential, commercial, institutional, recreational, industrial, transportation, etc., across the city. Corresponding strategies of regulation were enlisted in the form of "permissible use categories" to be overseen by the monitoring agencies. Confirmation of these regulations would ensure *a legal activity* on that particular piece of land. In the Master Plan, zoning was considered as "an emblem of urban health and banished clutter" (Sundaram 2010: 48). This categorization and distribution of the use zones could be seen as the "objectification" of the city. A complete faith on the homogeneous zoning method objectified the urbanity and was, perhaps, responsible for the plan's inability to understand the heterogeneity of activities in the older parts of the city, like Shahjahanabad and the "urban villages".

Nevertheless, a sense of proximity was attempted in the distribution of population, spaces, and activities in order to achieve equity of access. This assumption of the plan, I discuss later in this book, does not attain equity and justice in delivering and accessing resources.

Decentralization, hierarchy, and zoning, as Foucault (1975[1972]: 220) suggests, could be seen as the disciplinary methods of "partitioning and verticality" of separating different elements, thereby opposing the complex nature of multiplicity. For the convenience of control, the plan attempted various modules of the planning units with useful sizes.

Learning from the Master Plan 1962

I have focused this discussion on the MPD 1962 because it contained the original vision and policies, carried forward in the successive plans with certain modifications out of sheer necessities of the changing ground realties. The Master Plan has been an object of criticism from different point of views, some of which I have already mentioned. Having recognized the vital issues of those diverse critiques, I may now sum up some of the learning from the plan.

First, Master Plan was the production (or the apparatus) of the rationality of the policy of planning, approached from the ideology of welfare democracy of creating a city of *utopia*. But, the Plan became the *ensemble* delivered through the techniques of governmentality by corresponding with the mentalities (or the rationalities) of the policy of planning.

Second, Master Plan consciously tried to move away from the models of the caste-based community of Shahjahanabad or the rank-based colony of Lutyens' Delhi toward the option of a "secular" kind of neighborhood agglomeration through vaguely identifiable class formations by mixing the income groups and the size of properties. But, such a westernized version of neighborhood as a model of the "urban secular consciousness" has been considered a failure (Sundaram 2010: 58). However, the planners at the time of the Plan had aggregates of displaced people out of partition and individuals migrated for administrative jobs to the capital city. What kind of community could have been made is something to ponder about.[18]

Third, quantification and statistics, too, helped the plan to neutralize the qualitative prejudices or biases of religion, ethnicity, caste, etc. at a critical juncture of Indian history. The internalized desire of modern planning to shed off any preexisting notion helped the plan objectify the subject. Population and activities were then distributed over the spaces in cellular neighborhoods almost like social "melting pots."

Fourth, the distribution and the growth did not initially follow the "urban ecology model" or "class-wise spatial segregation," instead, it opted for the version of the "living mechanism analogy" of cell structure through two-way decentralized distributions of space and scale. This particular component of the plan is significant. However, later periods have seen the urban sprawl and the emergence of slums within the city and the subsequent resettlement of several of those slums at the periphery. In effect, a complex inconsistent pattern has evolved at an overall level. The notion of equity in the delivery of urban resources, like water supply, becomes even more important in such a mosaic of divergent living conditions. The difference and variation of social and urban fabric of the city can be seen even at a smaller scale-level. I refer this existence of complex inconsistency, which emerged outside the (policy) framework of planning, as the *real city*.

Fifth, the real city also consists of the present condition of the urban poor in Delhi, which, many feel, is due to the lack of implementation of the MPD 1962 and related bad governance. Shortfalls of housing provisions for the LIGs and somewhat indifferent attitude of urban planning and the government agencies toward them and their habitat have been held responsible for the state of the poor today (Bhan 2009, Baviskar 2002, Dupont 2004, 2008, Ghertner 2008, Kumar 2006, Roy 2000, Sharan 2006, Sundaram 2010, Tiwari 2003, Verma 2000).

Finally, the question is: Why are there so many deviations from the stated objectives of the Plan? What was delivered? How much and where? Some of the established critiques, mentioned before, answer many variations of these questions and there is a rich body of theoretical articulations of the *two-city* notions to explain the binary opposition between planned–unplanned, legal–illegal, authorized–unauthorized, formal–informal, planned colonies–slums, green city–brown patches, etc.

In the summary of the course on the "state formation" at the Collège de France in 1977–1978, Foucault identifies governmentality with "the manner in which the conduct of a mass of individuals comes to be implicated, in an increasingly marked manner, in the exercise of sovereign power" (Curtis 2002). He subsequently shifts the emphasis of the concept more toward the techniques of control and surveillance from the domination and the power of the sovereignty (Curtis 2002, Foucault 1978). Governmentality discourse, thus, tends to align itself more to an *administrative state* than to the *sovereign state*, and recognizes possibilities of unknown consequences at the delivery level of any policy decisions (Dean 1999). In other words, governmentality may expect that some policies would be delivered and others would not be, or a particular policy would be implemented in some situations but not in some other situations, or different policies would be formulated for different sets of people. Such observations on governmentality offer a framework for analysis. Several works find this discourse particularly useful in explaining the neoliberal governance that I shall revisit later (Barnett et al. 2008, Burchell et al. 1991, Corbridge et al. 2005, Dean 1999, Donzelot and Gordon 2008, Larner 2000, Legg 2005, 2007, Lemke 2001, 2002, 2007, Rose 1999).

As discussed, the Master Plan, delivered out of the policy of the state, has its rationalities (or mentalities) of distribution that had these generic techniques of governmentality built-in right from the beginning. Some of those rationalities did not even get implemented and delivered on ground to realize what I refer to as the *intended city* of the policies of planning.

The notion of the "poor welfare state" may have relevance in this context, but tendencies of the techniques of governmentality leading to the unknown consequences of *the politics of distribution effecting differential deliveries on ground*, may also be seen responsible for the condition of inequity and injustice.

Notes

1. Larner (2000) discusses Keynesian welfarism in establishing basic differences that neoliberalism has with it.
2. Kaviraj (2005: 13–14) feels "Nehruvianism" led to something different from socialism, but "it was not a failed socialist state."
3. There have been severe criticisms of the overall notion and its implementation in the act of nation-building in independent India (Chatterjee 1994, Kaviraj 2005, 2010, Kothari 1991, Thapar 1987).
4. Kaviraj (2010: 28–29 and 223–224) discusses these points critically. He also mentions that development in the Nehruvian State has been a loosely defined ever-expanding notion.
5. For the purpose of my discussions in this work, urban planning, city planning, town planning, and physical planning are considered analogous to one another without getting into finer disciplinary differences, if any.
6. There are a number of significant literature on the practices of the state policy and the decision-making processes pertaining to the specific planning and design of Chandigarh and Delhi, among which Scott's (1998) cultural–anthropological critique of Chandigarh and Sundaram's (2010) narrative of the historiography of the planning of Delhi are worth a mention here.
7. In 2006–2007, I was the coordinator of the faculty team that conducted an Urban Design Studio for the fourth year students of Architecture at Sushant School of Art and Architecture in Gurgaon, India, in which three towns in Haryana, namely Dharuhera, Rewari, and Bawal were taken up. It was astonishing to see the rampant urbanization planned by the HUDA for these small- and medium-scale towns. Interestingly, instruments and techniques of planning have remained similar to those of the MPD.
8. A glance at some of the significant critiques may help in recognizing the inadequacy of dominant urban planning and in aligning with the thematic courses of this work.

    Jane Jacobs in her book, *Death and Life of Great American Cities* (1964[1961]), gives the pioneering critique of both these rationalist ends of the city planning approaches. She mentions that "a city cannot be a work of art" and, hence, there is hardly any concurrence of the clean geometry and order of the city plans and the real aspects of day-to-day life (Jacobs 1961: 372–373). She brings out sociological issues due to simplistic distributions of single use zones in cities, which are: lack of "vitality" in terms of activities and social life of residential neighborhoods, spatial segregation of live–work activities and of social classes, and the "loss of street" as public place (Jacobs 1964[1961], Scott 1998, Taylor 1998). Her comments on the anti-urban attitude toward cities, inherent in the notion of the likes of Le Corbusier on the one hand and Ebnezer Howard on the other, brings out strong reservations about the overuse of sprawling open spaces full of "grass" and of the idyllic settings with less density (Jacobs 1964[1961]).

Another very significant criticism that breaks the very fundamental tenets of hierarchical distributions of the rationalist structure of the functionalist city comes from an empiricist understanding of the city by Christopher Alexander in his article, "A City is not a Tree" (1965). Alexander argues that constant overlapping of activities and spaces creates complex multiple realms of life in a city, the "semi-lattice." The "semi-lattice" of the spontaneous and evolving "natural" cities, he argues, gives richness to cities, as against the hierarchical "tree-like" distributions of facilities, functions, and components in the planned, "artificial" city.

Alan Colquohoun in his succinct discussions on representations of city and spaces in Corbusier's notions of urbanity refers it as "a kind of diagrammatic representation" of constituents of the modern city described by Georg Simmel in his significant text "Metropolis and Mental Life" (1950[1900]). In Simmel's understanding, all qualitative relations and their differences in modern city are abstracted to quantity (and to money) and Colquohoun argues that this notion of sociological abstraction is translated directly into the representations of "abstract geometrical form," order and spaces in the utopian grand city schemes of Corbusier.

Scott (1998) in his rather cultural-anthropological analysis refers to the utopia of grandness as the "high modernism" of the state practiced across the world in building the cities. "High modernism," he argues, operated within a particular "temporal and social context" having "uncritical, unskeptical and thus unscientifically optimistic" faith on "the legitimacy of science and technology" (Scott 1998: 4–7, 87–102). While including Chandigarh in his list of the actually built "high modernist city" alongside Brasilia, Canberra, Islamabad, Saint Petersburg, etc., Scott observes that most of these cities are the seat of administrative power of the state.

Scott's way of equating high-modernism of authoritarian planning with the authoritarian state is somewhat problematic especially in the context of cities like Chandigarh. One agrees upon the adoption of utopian high-modernist planning in Chandigarh, but its context was the democratic state of India. On the other hand, Colquohoun's arguments of the notion of representations of abstract geometry, form, and space in the utopian grand city schemes of Corbusier seem more relevant in Chandigarh as it was considered as the representation of independent, modern India. This understanding is also in tune with Lefebvre's (1991[1974]) concept of the *spaces of representations*.

David Harvey in one of his more recent articles (1997: 232) gives another critical perspective by differentiating the *process* (the urban) from the *thing* (the city):

The problem of these thinkers [like Ebnezer Howard or Le Corbusier] was not that they had totalizing visions or subscribed to master narratives or indulged in master planning. Their problem was not that they had a conception of the city or the social process as whole. Their problem was that they took this notion of thing and gave it power over the process.

He also problematizes a "dialectic" relationship between the notions of process (social, urban, political, etc.) and form (product, thing, political territories, city, country, etc.) in which process contributes to the creation of form, but once form is constituted, it tends to influence the process.

9. Foucault elaborates this notion of subjectification and emergence of bio-power in his book *History of Sexuality*, vol. 1 (1978).

10. Lemke (2001) does not necessarily agree with this observation of Balibar because it does not take into considerations Foucault's work on governmentality. Lemke, in fact, considers the governmentality critique of neoliberalism significantly close to some of the Marxist critiques of the same.

11. Balibar suggests this is where Foucault enters into a tactical alliance with Marxism:

    ... the first involving a global critique of Marxism as a "theory"; the second a partial usage of Marxist tenets or affirmations compatible with Marxism. ...Thus, in contradictory fashion, the opposition to Marxist "theory" grows deeper and deeper whilst the convergence of the analyses and concepts taken from Marx becomes more and more significant. (1992: 53)

12. New Delhi Municipal Committee was approved by the Chief Commissioner on March 16, 1927. New Delhi Municipal Committee was superseded in February 1980, and in May 1994 was renamed as the New Delhi Municipal Council (NDMC).

13. Later in continuation with the redistribution strategies, to avoid any large accumulation of land in the private hands, Urban Land (Ceiling and Regulation) Act was created in 1976 with parliamentary enactment, which was later repealed by most of the states realizing the limitations of the Act.

14. For example, if the present use of a particular plot of land does not conform to the prescribed land use by the Master Plan, there is a problem of legality. During the "sealing drive" in Delhi in 2006, such unauthorized use of land raised many legal, political, and social complications. In the context of the French society, an engineer, economist, and politician, Alain Lipietz (1980: 61) uses similar formulations in discussing territorial development based on the problematic that the "society recreates its space on the basis of a concrete space, always already provided, established in the past."

15. Foucault (1975[1972]: 167) discusses the notion of *tactics* as "the art of constructing, with located bodies, coded activities and trained aptitudes, mechanisms in which the product of the various forces is increased by their calculated combination are no doubt the highest form of disciplinary practice."

16. As per MPD 2021, National Capital Territory (NCT) of Delhi has a total area of 1,483 sq. km. Under MPD 2001 and MPD 2021, the NCT of Delhi is divided into 15 zones or divisions (A to P); out of which eight zones are in Urban Delhi (A to H), six are in Urban Extension and Rural Areas (J to N and P), and one is for the river and river front area (O).

17. The Master Plan 1962 also recognizes the fact that, in many cases, the planning division boundaries overlap municipal boundaries.

    ...to obtain workable units, the planning divisions have been further sub-divided into 136 development zones.... Change in land use, existing physical features, railway lines and major arteries act as boundaries for these zones. Municipal boundaries, election and census wards have also taken into considerations in drawing up these boundaries though they have not been a decisive factor in their delimitation. (DDA 1962: 64)

18. Perhaps, one clue for the community formation could have been addressed from the knowledge of the existing indigenous city. The Master Plan was, by and large, aloof toward the existing urbanism. Without an aptitude to learn from the existing urbanity, planners wanted to make it better—that was the mistake committed.

    Formulation of some strategies of intervention based on quantitative analysis and suggestions of certain out-of-place designs for qualitative enhancement were highly inadequate. Some of the criticisms of the quality of urban form and space that resulted out of the plan were also because of its (the Plan's) lack of engagement with the historicity. One may hold the design sensibility and the approach (or the lack of both) responsible for that.

4

Politics of Distribution

Now, I put forward the third proposition:

Governmentality refers to "different rationalities of the government" leading to techniques of distribution with in-built tendencies of social inequity and injustice. As a result, politics of distribution arises out of multiple practices of spatial selectivity causing differential delivery at the end.

Governmentality, I have discussed earlier, applies multiple techniques of normalization and possesses tendencies of "unpredictable consequences, effects and outcomes" (Dean 1999: 17–23). So, it is more inclined to be "selective" producing "uneven" and inequitable conditions. Such techniques, methods, and conditions are the outcome of what I refer to as *politics of distribution* (Figure 4.1).

The deviation between the *intended city* of the policy of urban planning and the actual delivery on ground to form the *real city*, then, could be seen in the selective delivery and distribution of resources and services or in the implementation of the policy. Also, when the state works across a time-span within different political and economic conditions, inherent unpredictable techniques of governmentality may accommodate corresponding deviations at the policy level as well. For example, in India, initial policies followed the ideology of the welfare state and its regulatory economy and since 1990s, neoliberal policies have contributed to the open, market-based economy.

Figure 4.1:
Politics of Distribution: Arguments and Conditions

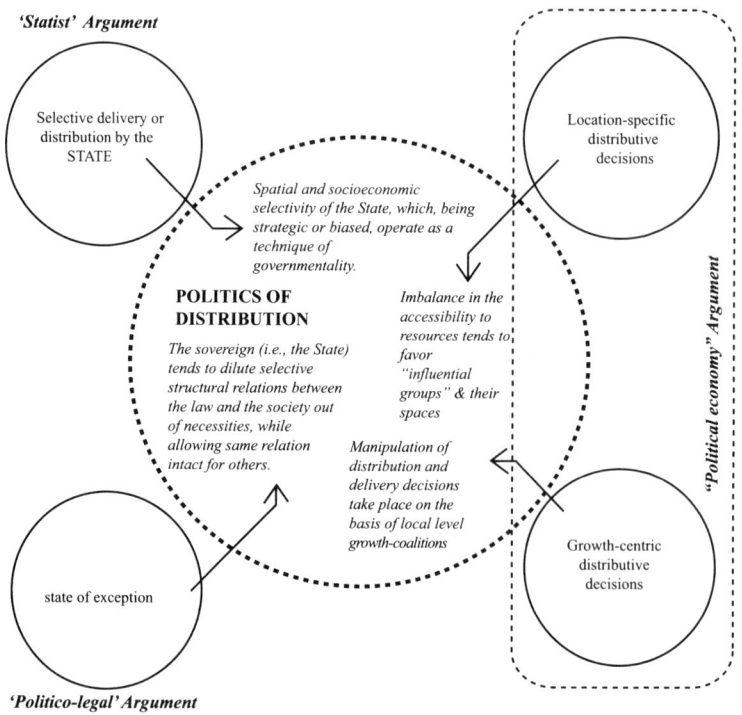

Source: Author.

The question is: How does the politics of distribution create conditions of inequity and injustice on ground? A tactical theoretical alliance of four different positions is attempted here to examine how politics of distribution takes place out of the multiple techniques of governmentality, which eventually causes inequity and injustice (Figure 4.1).[1]

State and Its Selective Delivery

I begin with some of the "strategic-relational" views on "the state theory" by Bob Jessop followed up by others (Brenner 2004, Jessop 1990, Jessop et al. 2008, Park 2008).[2] Discourses on the selectivity of the state often involve multiple scale-levels, like regional, state, or national level (Brenner 2000, Jessop et al. 2008, Smith 2008[1984]).[3] However, the application of the concept here is at the "urban" level.

Despite complex nature of the state apparatuses and institutions, Jessop (1990: 260–262), an eminent Political Theorist, observes a penchant of the state for favoring particular social forces, interests, and actors over others. state, then, is likely to be "a set of institutions that cannot, via structural ensemble, exercise power" and it has "no essential unity" in creating clear-cut limits for its institutions thereby leaving the latter in various conditions of ambiguity (ibid.: 117). State projects bring out state's character by adopting it both "as a site and an object of strategic elaborations" and "strategic selectivity" reveals the state's "bias" toward *specific strategies* to be adopted for *specific people*, to be implemented at *specific spaces* for *specific benefits* within a *specific time period* (Jessop 1990: 9, 10).[4] In such situations, state's selectivity becomes an outcome of "the evolving relationship between inherited state structures and emergent political strategies intended to harness state institutions toward particular socioeconomic projects" (Brenner 2004: 85, 87). For example, while formulating a policy, the state considers the social, economic, and political constituents as a whole, yet while implementing the policy, it selects a particular one or the set/s out of the whole constituents. This is somewhat similar to the totalization and the individualization of governmentality methods. I refer to it as the *socioeconomic selectivity*. Also, depending on where the state policies are actually delivered or implemented, the "spatial selectivity" occurs out of spatial advantages or benefits due to the differential delivery of policies (Brenner 2004: 90).

Under capitalism, with a definite role in the "production, regulation, and accumulation" of urban spaces, the state applies multiple strategies to bring hierarchy in social relations to maintain the "social cohesion" and the "functional distribution" over spaces and in doing so, the "state mode of production" utilizes, among other operations, physical planning and infrastructural investments over spaces (Lefebvre in Brenner 2000: 370–371).

Location-specific Distributive Decisions

One may perceive a possible link between the political-economic process and the city, especially the way Harvey (1997: 232) problematizes a "dialectic" relationship between the process (social, urban, political, etc.)

and the form (product, thing, political territories, city, country, etc.): the process contributes to the creation of form, and the form, once constituted, influences the process. Then, the notion of "urban" within a particular time and space comprises of a set of diverse processes which are social, ecological, political, and economic in nature and which form the city, and in turn, get enhanced by the city.

Harvey's approach is in continuation with a rich body of texts, the central argument of most of which situates the planning and its effect within the context of "political economy."[5] Lefebvre (1991[1974]: 8), for example, comments, "society as a whole continues in subjection to political practice—that is, to state power" and "all aspects, elements and moments of social practice" are included in "spatial practice." Lefebvre's (1991[1974]) Marxist phenomenology regards urban planning as an example where spatial practice incorporates social and political practices in constituting the urban reality.[6] If the city can be seen as an arena, "an understanding of which helps in understanding of the overall society which creates it," urban planning cannot be taken out of the social, economic, and political contexts where it is applied (Dear and Scott 1981, Harvey 1975[1973], 2003[1992], Pahl 1975[1970]: 234–235, cited in Taylor 1998: 102).

Let me introduce Harvey's discussions in his seminal work, *Social Justice and the City* (1975[1973]), on the notion of income in the perspective of "historical-geographical materialism." He begins with Titmus's definition of income:

> No concept of income can be really *equitable* that stops short of the comprehensive definition of which embraces all receipts which increase *an individual's command over a use of a society's scarce resources*—in other words, his net accretion of economic power between two points of time. ... Hence income is the algebraic sum of (1) the market value of rights exercised in consumption, and (2) the change in the value of the store of property rights between the beginning and the end of the period in question. (Titmus 1962: 34, quoted in Harvey 1975[1973]: 53; emphasis included)

Harvey (1975[1973]: 53) highlights two points from here: "income includes the change in the [real] value of an individual's property rights" which dictates "individual's command over resources;" and the command over society's scarce resource is also dependent on "the accessibility to and the price of these resources." He introduces the concept of "income charge" to take into account the combinations of earning (more or less), benefits (positive or negative), resource availability (more or less), and price of resources (higher or lower) (Harvey 1975[1973]: 53–54). Harvey (ibid.: 54) looks at how spatial form and social process in the city determine an individual's income. He identifies three aspects of spatial organization and political, social, and economic processes with the redistribution of income: location of job and housing, the value of property rights, and the price of resources to the consumer (ibid.: 86; Figure 4.2).

While underlining the aspects of the location of jobs and housing and its influence on the income distributions, Harvey tends to fall back on the "urban ecology" model of the American cities with the center of the city as the residence of lower income housing and the periphery being the suburban

Figure 4.2:
David Harvey's Political-economic Arguments on Distribution

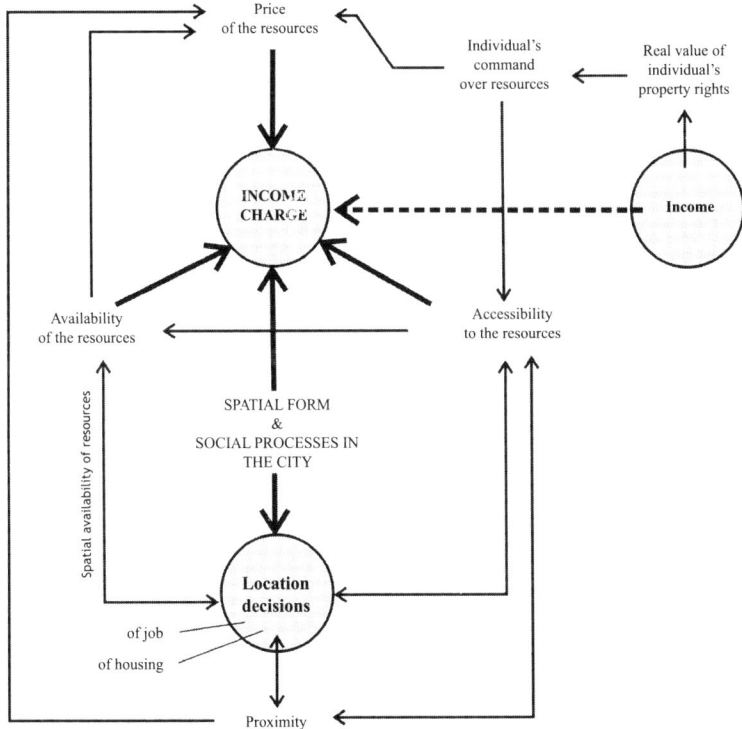

Source: Author.

high-end housing development (Harvey 1975[1973]). With that he ties up the notion of accessibility to avail of resources, workplace, or housing. Here, he considers the mono-functional land use distributions and with each land use, transportation time and cost become an issue (ibid.). When residential colonies inhabited by different socioeconomic groups coexist side by side, how shall the logic of distance work? We see such situations in most of the Indian cities, such as, Delhi, Mumbai, Kolkata, where high-end housing, slums, and LIG colonies jostle with each other primarily because of the urban transformations over a period of time in most of the cities in India. In the planning of Delhi, different income groups are represented over the territory of a planning zone within a larger decentralized model of distributions attempted by the MPD 1962. At the same time, urban transformations in Delhi have also seen, along with other instances, the relocation of slums from the middle of the city to the periphery and the urban sprawl of the *unauthorized para-legal* settlements at the outskirts by encroaching upon the reserved natural feature of the ridge.

Harvey (1975[1973]) also uses the notion of proximity to discuss the *value of property rights*. Certain favorable conditions, such as a park-facing or south-facing house in Kolkata and Delhi or a corner plot in a city, will increase the property value whereas unfavorable ones, such as noise, pollution, stink, will lead to the decrease of the value. A territorial organization of the city in terms of similar land values, utilities, and resources contributes to the social organization (ibid.). In India, there is a system of valuation of land by the respective state government, which determines the registration price of the sale or lease of properties. On similar lines, *property tax zones* have been identified in certain cities in India, such as Delhi, Bangalore, and recently in Kolkata, and colony-wise spatial categorizations of the city have been formulated for the collection of municipal taxes. Now the question is, when different territories are conceived or organized on the basis of land values or such similar criteria, does the distribution of utility and resources remain similar across similar territories over the city?

Harvey (ibid.) assumes that the spatial availability of the man-made resource is subject to location decisions for their distributions, and the proximity and accessibility, too, influence the local price of the resource. He puts forth an interesting point that the provision of resources from the perspective of the provider is different from that of the consumer.

I consider this particular point in the empirical research in the collection of data on water supply from the provider (the DJB) and the user/consumer (the people/household).

In an urban system, the "availability of and accessibility to resources" are tilted toward the benefit of the rich and the disadvantage of the poor, and such bipolar imbalance of distribution is often altered through urban politics with more powerful socioeconomic groups trying to "dominate locational decisions to their advantage" (Harvey 1975[1973]: 71, 75). Socioeconomic groups, therefore, sway the selective delivery decisions in favor of the spaces they live in; this situation may be termed as the *selectivity of the socioeconomic space.*[7] I address some of these issues by taking clues from Harvey's central assumption on how spatial form and social process in the city determine an individual's income and the question is: How do socioeconomic space (i.e., the spaces with certain income and social status) determine the delivery of urban resources in the city?

Growth-centric Distributive Decisions

David Harvey's notions, following the "urban ecology" perspective, comprehend the distribution of income, jobs, housing, etc., with relation to the spatial factors of locations, such as proximity, accessibility, transporting distance, availability of resource, etc. Alternatively, Harvey Molotch (1976) introduces, from the position of the urban political economy, the notion of "growth" that creates opportunities for jobs, housing qualities, utilities, business, etc., in specific localities of a city or some cities in a region or nation. City becomes "the human ecology" containing "maps of interest mosaics" (ibid.: 310). Urban political economy, to Molotch (1976: 31), is "the link between social organization and economic activity as mediated by earthly resources, particularly in the settlements where production, distribution, and consumption produce noticeable densities of human activity." His concept of "city as growth machines" was formulated initially by understanding the pattern of development in American cities and later has been discussed and developed in the context of other cities as well (Logan and Swanstrom 1990[2005], Molotch 1993, 1976). One can see the relevance of this discourse in recognizing the way in which cities have evolved or transformed over the years and even more in the recent neoliberal economic tendencies in countries like India. Also underlined is the contribution of the political participation and the influence to the opportunities resulting in different access by different social groups and locations (Molotch 1976). Consequently, a condition of the "growth coalition" in a particular area/locality gets formed (Mollenkopf 1992).

In this arrangement of coalitions, stakeholders having political and economic interests work toward maximizing economic development of that area and as a result, land rents of that locality increase with the chances of local growth (Molotch 1976). The formulation of the "growth coalition" is somewhat similar to Harvey's conceptions of the "influential community." Most of these so-called elites or businessmen make the political coalition of opportunities, mainly for status, visibility or, most crucially, for "making a living off the manipulation of place itself" (Molotch 1993: 31). In other words, the growth seems to be *interest driven* and not *location driven*, as the interest of the elites and local political and "hegemonic" business communities of one locality is in competition with that of the other locality. When the business is not mobile, such as real estate development, utilities, newspapers, etc. having specific ties to a place, growth has to create more opportunities: If the place cannot move to the people, people move toward the place.[8]

Molotch (1993: 313) observes that the organized effort by the government and the private interest groups in necessitating the growth distribution is the "essence of local government as a dynamic political force." Some of the local residents get involved in the politics, especially in the local party structure and fund raising, "for reasons of land business and related processes of resource distribution" (ibid.: 317). The political process out of some unseen negotiations with unpredictable consequences determines the material gains (by whom, what, where, and how) (Bordreau 2003, Edelman 1964, Fainstein 2005, Lasswell 1936, Molotch 1976).[9] This is the "politics of distribution" at the local level (Molotch 1976: 313–314). In this version of the politics of distribution, selectivity of the socioeconomic space is essentially interest driven.[10]

Politico-legal Notion of the State of Exception

Another interesting notion is by an Italian philosopher, Giorgio Agamben (2003[2005]), who brings "law" at the focus of discussion and exposes two sides of it: The objective legal framework and the subjective application in a state of exception. A state of exception forms a "point of imbalance between public law and political fact" within the ambiguous margin formed "at the intersection of the legal and the political" (Agamben 2003[2005]: 1; Figure 4.3).

Figure 4.3:
Relationship between the Sovereign, the Territory of Law, and a State of Exception

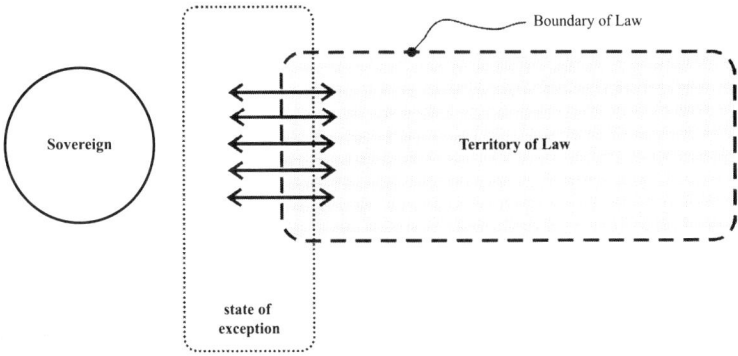

Source: Author.

Agamben's (2003[2005]: 24–25) central argument is: If the law guides the exception, it temporarily abandons itself for what can be seen as the necessity and since "necessity has no law," it makes its own law while being completely indifferent to any preexisting legal bindings; hence, the *state of exception becomes the "legal concept" of the condition of necessity.* "The sovereign who can decide on the state of exception guarantees its anchorage to the juridical order" and the decision, the sovereign (the state or the king or the dictator) takes, is guided by its ideology (ibid.: 35).

The sovereign, by being outside the law, can control others remaining inside of, what Foucault mentions as, the territory (Figure 4.3). Also by creating porosities in the legal framework, the sovereign makes provisions for a zone, the state of exception, to allow specific instances out of the compulsion of the established framework of law. A state of exception, thus, forms an uncertain gray area (of margin) beyond the binary oppositions of inside–outside, legal–illegal, etc.

The state of exception, while being selective in some situations, dilutes the structural relations of the law and the society, but at the same time, lets the same relation continue for others.

*   *   *   *   *   *

As indicated conceptually, planning gets manipulated by the politics of distribution. This process creates both spatial and socioeconomic conditions of selectivity, which are further connected with the legal selectivity in a state of exception.

In Delhi, urban planning is controlled by the state. The deviations between the intended city of the policy of urban planning and the real city on ground may very well be due to the politics of distribution, contributed by the selectivity of the state and further allowed by the inherent unpredictability of the techniques of governmentality.

Selectivity is not only handed top–down by the state, but also modified at the bottom and in both cases, policy is subverted. Politics of democracy in Rancière's (1999: xii, 102) central thesis has "the rationality of disagreement" and policy, the consensual practice of the government to the "conceptual legitimization of a democracy," becomes the condition of the "post-democracy." Planning, including urban planning, being the policy of the state happens to be post-democratic and, hence, planning is what politics is not. Chatterjee (1994, 2004), too, brings out the opposition between the conception of planning and that of politics with reference to the independent India. Planning policy, then, operates within, what Swyngedouw (2009) terms as, a "post-political" condition. It is obvious that the ideology of democracy did not have these notions of difference that the politics of democracy has (Chatterjee 1994, 2004, Rancière 1999, Swyngedouw 2009). Instead, ideology had the idea of equality and liberty contributing to the construct of the *idea city*.

It may also be possible to see the glimpses of the intended city and the real city existing together. Notions of such coexistences are somewhat loosely touched upon by few of the works I have mentioned here. Harvey's (1975[1973]) formulation, for example, has inherent in it an "urban ecology" model of spatial disposition involving location, income, and accessibility to resources. Traces of such spatial logic could be seen in many cities and in Delhi as well. But, cities are in an intermediate process when there is an assemblage of spaces which are *intended* in the model (the planning model) and the ones that have already departed the model. Mototch's (1976, 1993) notion of the city containing the "interest mosaics" of people also indicates the multiplicity of human existences.

Lefebvre (1991[1974]) considers that the representation of space in planning projects illustrates the utopia of a possible future within the framework of the real and the existing mode of production. Hence, the simultaneity of the real and utopian spaces seems possible. All these ideas are "uncomfortably" close to Foucault's notion of heterotopias, which talks about the coexistence of the real and the imaginary almost like a mirror. A mirror constructs a situation in which the reflection of the body and the body itself coexist.

In the course of theoretical formulations, several key factors influencing the selective delivery of resources and services have emerged: The first relates to the notion of legality of land that is *property* (and the quasi-presence and complete absence of legality); the second connects to *socioeconomic groups and urban spaces* they inhabit; and the third is concerned primarily with the *political patronage* influencing the delivery.

Now, the question is: What kind of urban services are to be delivered that can be the central point of my discussions from here onward?

I consider the delivery of water supply in Delhi as the entry point to the empirical part of the research. The empirical work illustrates, on the one hand, the broad territorial variations in water supply at the city level, thereby confirming the importance of spatial distribution and, on the other hand, the difference in the supply at the local level across socioeconomic strata, underlining the issue of opportunity in accessing the resource. Water is the prime "need" for life and, therefore, it is to be distributed to "make contribution to common good." Water supply in Delhi is under the control of the state; hence, the ideological obligation of the government (in other words, "the merit good argument") to supply water to all cross sections of the society comes to the fore. Also, water is a scarce resource in Delhi. The condition of equity and justice in the delivery of water supply, then, becomes a key issue and requires a close introspection.

Notes

1. Roy's (2003, 2005, 2008, 2009) notion of "informality" suggests a tendency of top-down decision-making by the government that is, often, in contradiction with the planning policy. On the other hand, Appadurai's (2001) idea of "deep democracy" and Holston's (1998, 2008) concept of "insurgence," both, to an extent, suggest activities initiated at the

"bottom," where the state policy is delivered on ground. These theoretical formulations, albeit differently, indicate the process of subversion and opposition happening in reality to limit the blanket policy implementation, and, are worth a mention here.

2. Most of these later discourses on the selectivity of state are in the backdrop of capitalism in the time of globalization, whereas some of the preceding Marxist texts sharing political economy perspectives are the critiques of industrial capitalism (Harvey 1975[1973], Lefebvre 1991[1974], Molotch 1976, and others). Yet, in the context of this book, one sees strong resemblance between both sets of work.

3. Jessop et al. (2008), for example, establishes the "territory-place-scale-network" theoretical schema to bring together various concepts involving the space and the scale.

4. Jessop draws heavily on Nicos Poulantzas's version of the "state as a social relation." He discusses at length a comparative analysis on the approaches of Poulantzas and Foucault on strategy and power, based on their respective works in the 1970s (Jessop 1990: 220–241). Poulantzas, identified as a structuralist Marxist, takes clues from Foucault's work. But, he deviates from the Foucauldian notion of the "bio-power" and disciplinary techniques and develops a view of the "state power as a social relation which is reproduced in and through the interaction between the institutional form of the state and the changing character of the political class forces" (ibid.: 221).

5. In the 1970s, a series of writings turned the attention of planning critique toward the social, economic, and political aspects of urbanism (Castells 1977, Dear and Scott 1981, Harvey 1975[1973], Lefebvre 1991[1974], Molotch,1976, Smith 2008[1984], and others). Most of these works, partly or completely, have been well documented in many edited compilations of city-related texts in the recent past (Bridge and Watson 2002, Cuthbert 2003, Kelbaugh and McCullough 2008, Kleniewski 2005, LeGates and Stout 2003[1996], Miles et al. 2000, Stein 1995).

6. It may be emphasized that along with Harvey, other stalwarts, such as, Henri Lefebvre, Neil Smith, and Edward Soja have been active in bringing *space* at the center of the discourse of political economy and the politics of distribution. Lefebvre's path-breaking work, *The Production of Space* (1991[1974]), on conceptions and practices of space is worth a special mention. The space of society as social spatiality is seen as simultaneously lived, conceived, and perceived, or, as he explains, as "material Spatial Practices, evocative and imaginative Representations of Space, and the complex, combinatorial, and never fully knowable Spaces of Representation" (Lefebvre 1991[1974]: 356–357, Soja 2001). Spatial

practice is the common interface where the social and political practices overlap (Lefebvre 1991[1974]).

7. In search of better quality of life, similar group of people can also move out of the spaces where they used to reside. In this instance, socioeconomic groups are selective about spaces where they live and as a result, the urban restructuring processes, such as gentrification and suburbanization, are initiated.

8. Let me cite few examples to illustrate the point.

    Due to the political agitations by the major opposition party in Singur in the state of West Bengal, when the Tata Motors was looking for an option for moving out the Nano car production beyond West Bengal, we could see how the state governments of Gujarat, Uttaranchal, and Karnataka were competing with each other by offering more facilities, promises, and concessions to the automobile company for shifting their operations. Finally, Tata Motors shifted the production of the Nano car to Gujarat. It was politically symbolic as well: The iconic automobile production left the state, ruled by the known Lefts, to go to the state, ruled by the known Rights. This is an example for *competition for growth opportunities* between states/regions.

    It can happen in another scale and sphere too. Anyone from Kolkata can vouch for the fact that even before a set of awards was introduced a couple of decades back by a paint company, localities unofficially used to compete with one another for drawing bigger crowds for their respective *Durgapuja* festival. In doing so, *puja* committees are required to arrange money through donations and advertisements which cannot be done without the involvement of local politicians, property developers, or businessmen. In today's political economy, these are the so-called elites and one can see the profusion of them in most of the organizing committees of the bigger *pujas* in Kolkata. It is an instance of forming the *social coalition* at the locality level.

    On the other hand, at a very petty level of distribution, one can see many examples of collusions. Some of my acquaintances in Delhi, who happen to stay in the same colony of a local politician, like a Member of Legislative Assembly (MLA), mentioned few times that there was hardly any power-cuts in their locality. This was at a time of regular power-cuts in the surrounding areas, in one of which my residence was located. It exemplifies the manipulation at the very lowest level of distribution.

    These three instances and many more unsaid ones that each one of us encounters quite frequently illustrate the influence of the interest-driven politics of distribution.

9. Young (2011[1990]: 39–45), refers to power relations as well: "who benefits from whom and who is dispensable" when oppression is faced

by social groups. Her notion of the "dominant group" is a broad-brush representation of Mollenkopf's influential group at a local level.

10. In the late 1970s, Oldenburg (1978[1976]), a political scientist, cites in his work, *Big City Government in India*, case studies of municipal wards in the older part of Delhi to argue that *urban politics* is an important part of local level governance in Indian democracy.

PART II

Politics of Distribution in Water Supply

5

Physical Setting and the Method of Study

It is necessary now to observe the interrelationships between the indicators, like planning, politics, service delivery, and socioeconomic spaces, in order to comprehend the *just* distribution of urban basic services and the *social justice* in accessing the service delivery. Theoretical positions and propositions bring out certain specific variables in relation to the selectivity in the delivery of services in urban conditions, which are: *property* as the legal notion of land, *socioeconomic groups and spaces* occupying the city, and the *political patronage* playing a role in the delivery of urban services.

The delivery of water supply in Delhi is considered as the key aspect to observe and understand, empirically, the spatial and social selectivity in the distribution of resources in the city. The central argument is: Even under the control of the state, the delivery of water supply in Delhi varies with the socioeconomic groups and spaces in the city.

Objectives of the empirical work are essentially threefold: First, to figure out aspects of governmentality in the delivery of urban utility services, like water supply, planning and politics, and their interrelationships in comprehending the equitable distribution; second, to understand real distributional patterns by identifying and analyzing links between the delivery of water supply and the socioeconomic spaces in the city; and third, to find out tendencies in political patronage in the operation and maintenance of water supply services across different socioeconomic spaces in the city.

The selection of water supply as the key empirical variable is loaded with a couple of assumptions. First, water is the prime need for life and sustenance and, hence, its distribution makes contribution to common good. Second, planning, being inclusive of the social, physical, and economic networks and processes, must recognize, contribute, and improve the distribution and delivery of the infrastructure in shaping the city. Not surprisingly, Harvey (1975[1973]: 100), too, identifies three criteria of distribution for achieving *social justice* as need, contribution to common good, and merit.

I will begin with the second point. There is an apparent tendency to assume technological reasons behind any inequitable distribution of infrastructure, whereas many believe otherwise (Graham and Marvin 1996, Mason 2002, Mitchell 1999, Pope 2008). Interrelationships between social and physical networks of the city acknowledge infrastructure as something "more than just the Pipes" and as much a social process as a technological one, where technologies cannot be isolated from the political and social processes managing them (Mason 2002, Pope 2008). Perhaps, a key issue is that "[t]he systematic disassembly of a mass society by the consumer economy is nowhere more evident than in the recent transformations of the infrastructural form" (Pope 2008: 18). In other words, the networks (and the physical layouts) of the infrastructure services become an important ingredient to the changing paradigms of both the social existence and the urban structure. Toward the end of this book, I shall return to some of the issues and concepts regarding the relationship that the urban structure may have with infrastructure distribution and delivery. The concern, at the moment, is to explore the connection, urban infrastructure, and its delivery share with the individual subject and the society at large as well as with the planning of the city.

One may now consider the broad classification of urban infrastructure for further explorations. Contemporary urbanism and the uneven connection of globalization, both, are partly constituted, supported, and driven by drastic transformations of networked infrastructures, mainly telecommunications, transport, energy, water, and urban streets, which mediates the process of modification of nature into the city (Graham and Marvin 2001, Sassen 2000[1994], Schirato and Webb 2003[2006], Smith 2008[1984]). For example, water is sourced from a distance, then treated, stored, and supplied to wide variety of individuals and groups through complex and often invisible mechanisms. Such dynamics of the socio-technical process of exchange guides the construction and the use of the networked infrastructure in the city (Graham and Marvin 2001). The notion of the globalized networked society is obviously ingrained into these arguments and, therefore, the question of unevenness cuts across spatial boundaries. On the other hand, within a particular area of a city, where technology (of the delivery of such infrastructure) may not change so much across the socioeconomic groups and their spaces, yet the actual delivery, I argue, varies across those groups/spaces and causes inequitable conditions.

Till today, wherever the government controls infrastructure, the equity and justice in the distribution and delivery is expected. Such an involvement of the government in the supply of goods and services, in general, is also used for broader classifications of infrastructure. Batley (1996) identifies three arguments for the government intervention in the infrastructure delivery even within the "market-driven private operations", the public good argument, market failure argument, and the merit good argument (ibid.: 726, 728). "The public good argument" includes non-competitive and collectively beneficial basic services, such as, street sweeping, street lighting, etc., for which individual consumption is difficult to measure (ibid.: 726–728). "The market failure argument" illustrates instances of inefficiency in "market-driven private operations", for example, monopoly of services like water supply, or big capital investment with uncertain returns, like major underground networked infrastructure, or "positive externalities" by offering services, like vaccination, to unwilling consumers, and so on (ibid.). "The merit good argument" is the ideological obligation of the government to provide certain goods and services, such as education and health, to all cross sections of society (ibid.). Based on these arguments, a broad classification of public services can be identified as: the "urban infrastructural services," such as water supply and sewerage, the "infrastructural development," such as infrastructure installation and slum improvement, and the "personal or household services," such as primary education and solid waste management (Batley 1996: 729).[1] Located within a discussion of possible public–private partnership (PPP) under neoliberal economic scenario, this classification attends to the user-end of the society and, simultaneously, addresses the ideological inclination of the state.

Planning, as elaborated earlier, was conceived in the post-independence India as a policy following the ideology of the democratic welfare state. Till the end of the 1980s, the state was almost the sole provider of basic goods and services to the people for the betterment of their quality of life. Since the 1990s, with the process of globalization and economic liberalization, the urban governance has significantly started moving toward the privatized operations, often under the PPP mode, in many sectors including the provision of several infrastructure and basic services. The inefficiency in government functioning has been the prima facie logic behind such operational maneuvers.

Delhi was the first city comprehensively planned in the independent India and till today, the operation and maintenance of water supply, essentially considered as a *public good*, is controlled by the DJB, a public body under Government of the National Capital Territory of Delhi (GNCTD). Ideologically, the state would have conceived a uniform access to a pubic good, like urban water supply, across all sections of the society. But at the policy level, per capita allocation of water in Delhi is dependent on the land tenure status or the settlement types akin to the suggestions made in the Master Plan (Central Public Health and Environmental Engineering Organization [CPHEEO] 1999, Hoyt et al. 2005, Maria 2008, Singh and Shukla 2005). For example, norms for water supply in planned colonies are 225 liter per capita per day (lpcd) as compared to a meager 70 lpcd for the slums (Table 7.3). The legal notion of property at the settlement (or colony) level seems to have a significant input to the policy-framing of the state, in general, and water supply, in particular.[2] The policy of allocation of water supply with respect to the settlement types and its respective land tenure makes one apprehensive about the state of equity and justice in actual delivery at the end.

The fundamental assumption, here, is that the water supply varies with the several gradations (or shades) of socioeconomic stratifications existing even within the legal property, of which planned colonies are also one part. Accordingly, multiple levels of delivery are created at the end. In place of a binary construct, the city is read as a space with multiple realties and conditions, to use the Foucauldian term, a "heterotopia" (Foucault 1984[1967]). These conditions are realized even when one introspects into the equity and justice of the delivery of water supply. How does the delivery of urban water supply in Delhi correlate with city's socioeconomic spaces and its legal property rights?

One then gets curious to know the correspondence between legal property rights and socioeconomic space. In the empirical research here, the notion of property is used at the settlement or at the colony level. The legality of an individual property is not of concern in this context. The term, legal property, here, indicates various rights, length of time and the manner of holding the land, and the land tenure, which is approved by legal instruments. The planned property, built as per the Master Plan, is part of the legal property. However, the right to live, especially for the lowest sections of the society, in a city, like Delhi, is a complex arrangement out of multiple factors. The view that the right to live cannot exist outside the legal purview would be rather simplistic and I will touch upon this discussion elsewhere in this research by drawing on several significant works (Bhan 2009, Chatterjee 2004, Holston 1998, 2008, Holston and Appadurai 1996, Marshall 1950). In this work, mainly colonies (spaces) with legal tenure of land are to be investigated. Such colonies are covered under the property tax categories of the MCD. I use property tax categories of the MCD to make empirical observations on diverse socioeconomic spaces of the residential setting of Delhi (Figure 5.1). In the absence of any direct measure of the socioeconomic

Figure 5.1:
Empirical Research Variables and Its Interrelationships in Real Distribution

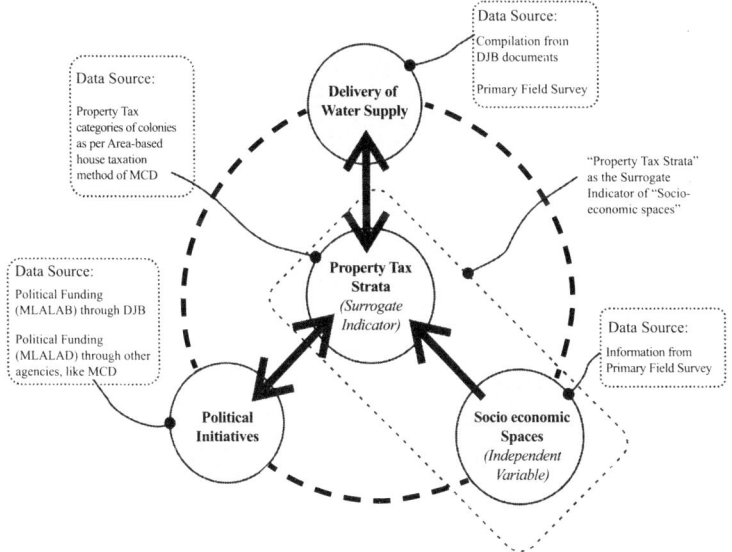

Source: Author.

characteristics, property tax would eventually become its appropriate surrogate indicator of classifications and can be used as an independent variable for the statistical analysis of empirical data.

In case, multiple practices of delivery of water supply exist in Delhi, the implementation is likely to differ from the policy and the policy from the original ideology of the welfare state. Governmentality forms the wider theoretical source of such practices of delivery, which tends to drift away from the policy (of planning) toward the sphere of politics (Chatterjee 1994, 2004, Rancière 1999). The delivery practices, quite often, are manipulated by the "political patronage" out of the coalition between the politicians and the other influential and powerful residents or, the "patron -client relationships" based on the vote-bank politics (Chatterjee 2004, Chakrabarty 1989, Harvey 1975[1973], Lasswell 1936, Logan and Swanstrom 2005[1990], Mollenkopf 1992, Molotch 1976, 1993, Scott 1972). All these possible arrangements of patronages through many unseen negotiations can influence the implementation, either in favor of the richer or the poorer sections of the society.

Now, the question is: How does political patronage relate to the socioeconomic spaces (indicated by the property tax strata) in Delhi? More specifically, one may also find out whether the spending of the political funds favor *the spaces*, occupied by either the poorer or the richer socioeconomic strata, or are the political funds selectively spent around *the time* of the election. I address this point through observations on the spending of the political fund. The discretionary political funding, working outside the purview of the regular decision-making process of the executive, would expect to bring about a "patron–client relationship between the people and their elected leaders" (Kumar 2005). However, practices across all spatial levels, I argue here, have multiple governmentality conditions and may not overwhelmingly favor either the rich or the poor. Yet, one may observe an overall trend in governmentality practices that the richer socioeconomic groups are more privileged, even amongst the so-called *legal* owners or citizens for whom policy norms seem to be same.

Five broad criteria are identified for the physical setting of the empirical study. First, the area should be within the similar gross average water supply so that, to start with, not much of a difference in the overall supply across the spatial extent may be assumed. Second, it should be an area formed by the MPD so that the planning policy and its recommendations are in place. Third, it should have a relatively large size so that smaller areas at various locations, relatively away from one another, may be taken for analysis. Fourth, the area should include all scale-levels of political, planning, and infrastructural (water supply) jurisdictions, so that territorial overlaps and convergence can be addressed. Finally, it should comprise of diverse socioeconomic groups so that the key issue of equity and justice across various groups can be addressed. South Delhi is one such area and the setting of this empirical work.[3]

Inequity in water supply occurs at the broad spatial levels across Delhi, an issue that I will elaborate later (NCRPB 1999, PWC et al. 2004, Ruet et al. 2002, Susheela et al. 1996). Based on an average level of water supply, the NCRPB prepared (1999) indicative maps showing existence of several such spatial extents/areas having different gross average supply (Figure 7.5). South Delhi has been indicated in the map of the NCRPB as one of such area. Lower level of supply (148 lpcd) in South Delhi as per the NCRPB (1999) also makes water a scarce resource, considering the high demand from a largely medium/high-income residential area. A comparison of areas across the city may, then, be faulty because of the overall difference in the supply across such areas. Majority of the area in South Delhi is also covered within the South I and II of the DJB operational zones (OZ), which is expected to have similar technical dimensions of networks and installations across its areas. Since areas in South Delhi are provided with similar gross average water supply serviced within a given technical setting, differences in the overall supply is expected to be minimal across its territorial extent.

South Delhi, primarily a creation of the post-independence planning process, should be a suitable example to discuss the policy and recommendations of successive plan documents. A large part of South Delhi per se falls within the Planning Division (zone) F of the Master Plan (1962, 2001, 2021). F zone, divided into 19 subzones (F-1 to 19), covers an area of 119.58 sq. km, which is 8 per cent of the total area of Delhi. Because of the large physical territory of the zone, smaller areas can be taken up for analysis at various locations relatively away from one another. The zone accommodated about 15.75 percent of the total population of Delhi in 2001, witnessed, between the period of MPD 1962 and MPD 2001, an increase of about 34 percent of the population than the expected holding capacity in 2001, and is envisaged to accommodate 15 percent growth by 2021 (Table 5.1; DDA 2007a: 10). Also, F zone is the only one among three zones (A, E, and F), having maximum differences between the actual population and the holding capacity, that is to accommodate further population growth by 2021 (Table 5.1). The population growth

Table 5.1:
Estimated Holding Capacity of Existing Planning Zones (Population in Thousands)

Zone[a]	Holding Capacity MPD 2001[a]	Existing Population 2001[a]	Holding Capacity MPD 2021[a]	Difference in Population Projection	Anticipated Population Growth in 2021
	X	Y	Z	(Y–X)/X	(Z–Y)/Y
A	420	570	570	36%	0
B	630	624	630	–1%	1%
C	751	679	788	–10%	16%
D	755	587	813	–22%	39%
E	1,789	2,798	2,800	56%	0%
F	1,278	1,717	1,975	34%	15%
G	1,490	1,629	1,955	9%	20%
H	1,869	1,226	1,865	–34%	52%
Subtotal	8,978	9,830	11,400	9%	16%
Dwarka		597	1,700		185%
Rohini III		96	160		67%
Rohini IV and V		198	820		314%
Narela		179	1,220		582%
Subtotal	3,222	1,070	3,900	–67%	264%
Total	122 lakh	109 lakh	153 lakh	–11%	40%

Source: Compiled by the author.
Notes: As per DDA (2007a: 10), population figures are centered around "broad planning guidelines" and remaining population projected for the year 2021 is to be housed in newly "planned urban extensions."
[a]DDA (2007a: 10).

of F zone is upward, widening the gap between the census figure and the projected one over the years (Figure 5.2). Resources in this zone were obviously required to serve more than the estimated population. The distribution of *scarce* resources, like water supply, therefore, becomes a phenomenon of curiosity!

F zone is broadly monofunctional and in its overall land use structure, to an extent, a microcosm of the larger city. The land use composition of F zone is not expected to change substantially in 2021, yet it will further house more than 2.5 lakhs of people by 2021 (Table 5.2). A significant number of residential colonies may undergo the re-densification process and the Draft ZDP 2021, too, identifies such areas for redevelopment, especially the low-density government housing precincts (DDA 2007b). MPD 2021 specifies a number of planning measures for F zone: Local area planning, redevelopment of villages, unauthorized colonies, and

Figure 5.2:
Decennial Population Growth in F Zone

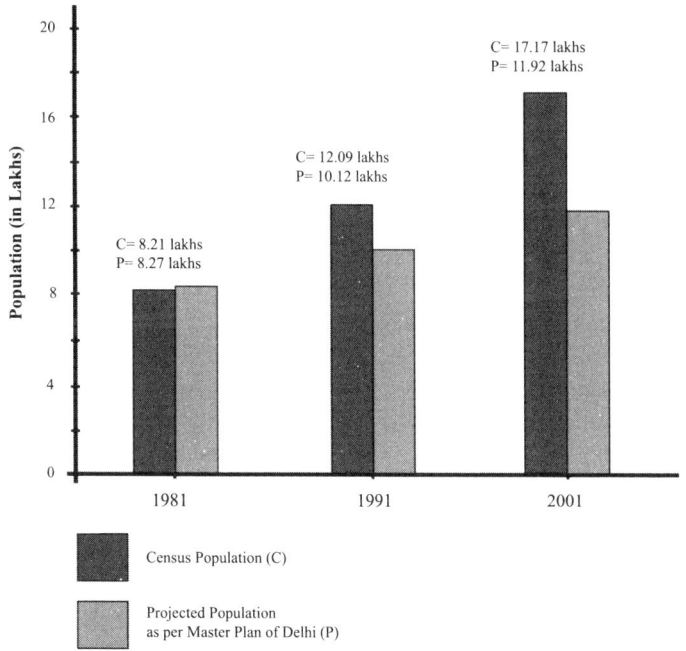

Source: Author.
Note: Data source is DDA (1998, 2007a, and 2007b).

Table 5.2:
A Comparison of Major Land Use Composition: NCT Delhi and F Zone

	NCT Delhi	F Zone	
Major Land Uses	As per MPD 2021 (%)	As per Draft ZDP 2021 (%)	As per ZDP 2001 (%)
Residential	45–55	35.43	35.60
Commercial	4–5	1.91	2.20
Industrial	4–5	4.29	4.30
Green/Recreational	15–20	26.53	25.90
Public/Semi-public	8–10	11.50	11.50
Circulation/transportation	10–12	10.78	10.94
Govt. office	Not Available	0.51	0.51
Govt. office undetermined	Not Available	6.80	6.80
Utility	Not Available	2.25	2.25
Total	100	100	100

Source: Compiled by the author.
Note: Data source is DDA (1998, 2007a, and 2007b).

built-up areas; restructuring and upgradation of residential areas in the influence zone of the Metro rail corridor and other major transport corridor; and incorporation of the network of recreational and sports facilities in detailed schemes.

F zone is controlled by the MCD and contains all scale-levels of political jurisdictions, water supply zones, and colonies with all urban variants of property tax categories of the MCD.[4] DJB is responsible for water resources management, monitoring the pollution of water, and the treatment and supply of potable water, wastewater collection, conveyance, treatment, and disposal facilities in the MCD areas (PWC et al. 2004). South Delhi water supply zones, by and large, fall within the planning zone F. Each supply zone (like Greater Kailash I, Jal Sadan, and the rest) has a number of subzones. Each subzone, then, has pumping stations supplying water to nearby residential colonies. However, there are hardly any maps or drawings available in the DJB to accurately delineate these zones on a detailed map of Delhi or at the levels of any of its planning zones (PWC et al. 2004). Nevertheless, the MPD is expected to plan an equitable distribution for basic civic services, like water supply, across at least one planning zone, if not for the whole city. The zone includes all three scale-levels of political jurisdictions, Parliament constituencies (PCs), ACs, and MW, with representation of different political parties at various levels. Political constituencies are important jurisdictions because the overlapping of political territories of the government and the role of ruling parties are key facets of the coordination between the administration and the governance. Political differences and conflicts surfaced quite often between the governments of Delhi state and the MCD, ruled by two different parties (Jain 2009, Nallathiga 2008, Pinto 2000, Sarkar 2009). The tension of the governance is, then, the result of administering over almost the same territory (Kundu 2006, Siddiqui et al. 2004). Also, the MLAs are found interested in using the MLALAD fund to bring to unauthorized colonies certain basic facilities, like the delivery of water supply by the DJB, which are otherwise outside the purview of the regular institutional provisions (Tawa-Lama Rewal 2005). When the Member of Parliament Local Area Development (MPLAD) spending on water supply is rare and the Councilor fund cannot be spent through a para-statal agency, like the DJB, the MLALAD is the most important local area development fund with respect to water supply.

Across one political constituency (Parliament/Assembly/MW), political initiatives in the provision of basic civic services, like water supply, are expected to be same. The representatives, the MP or the MLA, or the Ward Councilor, from the same party are likely to have similar interests in their political initiatives as well. Situated within an existing three-tier system of government in Delhi, party-politics is a relevant factor when different parties rule at different levels. Many processes may involve political initiatives in the delivery of water supply, but the most direct ones are the local area development funds, allowing the legislators to select any infrastructural project in their respective constituencies.

The discretionary political funding encourages a "patron–client relationship between the people and their elected leaders" (Kumar 2005).[5] The spending of the local area development fund would then consider the poorer section of the society as the *clients*, who otherwise receive limited infrastructural provisions from the procedures of governmentality (Chatterjee 2004).

Figure 5.3:

Examples of Urban Tissues in Some of the Residential Colonies in F Zone

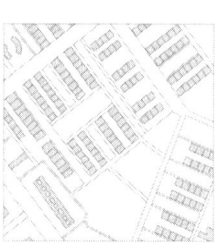

Maharani Bagh: Plotted residential colony
Property Tax Category: B
    Mostly 500 square yards plots with three-storied buildings with large open spaces at the community levels. Coarse urban grain with uniform, homogenous texture.

Sriniwas Puri: Govt. Housing
Property Tax Category: D
    Mostly, two-storied flats with (MIG/LIG dwelling units) with linear blocks sparsely placed around open spaces of varying scale and nature. Coarse urban grain with uniform, homogenous texture.

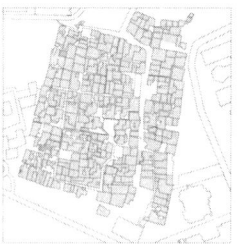

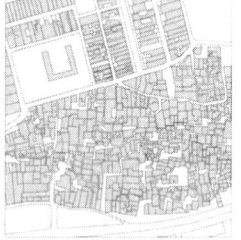

Jia Sarai: Urban Village
Property Tax Category: E
    Initially one or two-storied courtyard type residences that have been transformed in height and usage over the years. Irregular plot-sizes and built-to-edge of the narrow winding streets are the features of these areas. Fine urban grain with uniform, homogenous texture.

Hauz Rani: Urban Village
Property Tax category: G
    Basic morphology is similar as Jia Sarai with more dense fabric and quite often smaller plot sizes. Fine urban grain with uniform, homogenous texture.

Source: Author.
Note: Author's analysis is based on the map prepared from the images. Available at: http://www.earth.google.com (accessed on March 20, 2014). Each of the above Urban Tissues comprises an area of 250 meters × 250 meters.

F zone has the mosaic of urban forms with a generic urban character of low density and green areas. Along with the planned and well maintained posh residential colonies, the zone includes rehabilitation colonies, government housing areas and urban villages, and accommodates important heritage precincts of Delhi (Figure 5.3). South zone of the MCD, which falls within the large part of the F zone, comprises residential colonies across all property tax categories (A to H), which points toward the presence of diverse socioeconomic groups in this area.

The distribution and delivery in planning and basic services usually operates within certain spatial constructs, such as territories, zones, and locations. As maps are popularly used as an investigative spatial tool one potent activity can be the spatial mapping in identifying the territorial extents of the empirical work. The delineation of the study area has been one such kind of spatial analysis (Figures 5.4–5.10). Separate layers of political constituencies (namely Parliament, Assembly, and MW), planning jurisdictions, and their convergence are mapped. Infrastructural, for example, water supply, jurisdictions are also conceptually looked into. In this way, the territorial control and the overlapping of various agencies can be understood. Also, the property tax categories of the colonies are mapped spatially and the clusters of different property tax categories are observed for the selection of specific colonies for primary survey.

Accordingly, residential colonies may be chosen from these sets to observe variations in services across one particular jurisdiction.

Primary household survey was conducted in two phases over different parts of the study area during the period of 2006–2007 (Table 5.3).[6] The objective was to find out water supply delivery patterns over various colonies and across different property tax strata (socioeconomic spaces), and the target population included the users or the recipients of services in certain spatial units selected in South Delhi. Phase I of the household survey was carried out in selected clusters of colonies in the Okhla AC, whereas Phase II was over the same within a much larger area covering three ACs, namely Kasturba Nagar, Hauz Khas, and Malviya Nagar.

A residential colony was considered as a basic spatial unit for analysis. Different groupings of such colonies were compared and analyzed to

Figure 5.4:
Conceptual Mapping for the Selection of Hypothetical Study Areas (or Colonies)

Set 1. When all jurisdictions remain the same	Set 2: When all jurisdictions vary

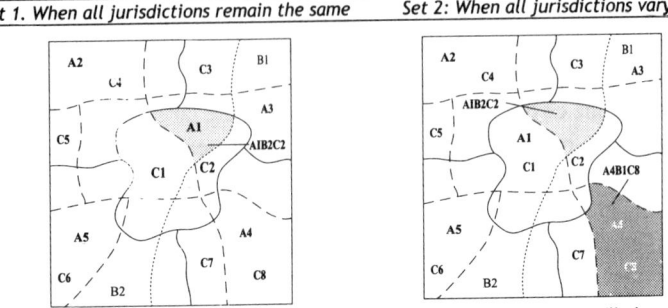

Residential colonies will then be chosen from A1B2C2 or similar other areas for observing whether services vary across that particular area or remain similar.

Residential colonies will then be chosen from A1B2C2, A4B1C8 or, similar such areas for observing whether services vary across that particular area or remain similar.

Set 3. When one of the jurisdictions (political or planning or services) remain constant

3a.   3b.   3c.

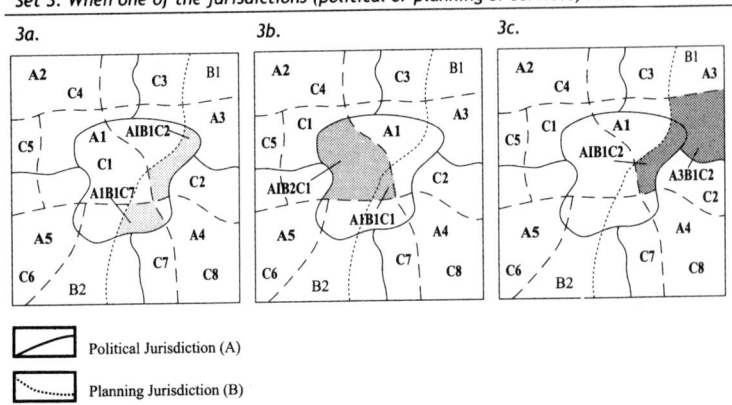

☐ Political Jurisdiction (A)

☐ Planning Jurisdiction (B)

☐ Infrastructural Jurisdiction (C)

Source: Author.

- Set 3a is to be used for comparing services in different infrastructural jurisdictions across same political and planning jurisdictions: for example, A1B1C2 and A1B1C7
- Set 3b is to be used for comparing services across different planning zones within same political and infrastructural jurisdictions; for example, A1B2C1 and A1B1C1
- Set 3c is to be used for comparing services in different political jurisdictions across same infrastructural and planning jurisdictions; for example, A1B1C2 and A3B1C2

Figure 5.5:
F Zone and Subzones in Delhi

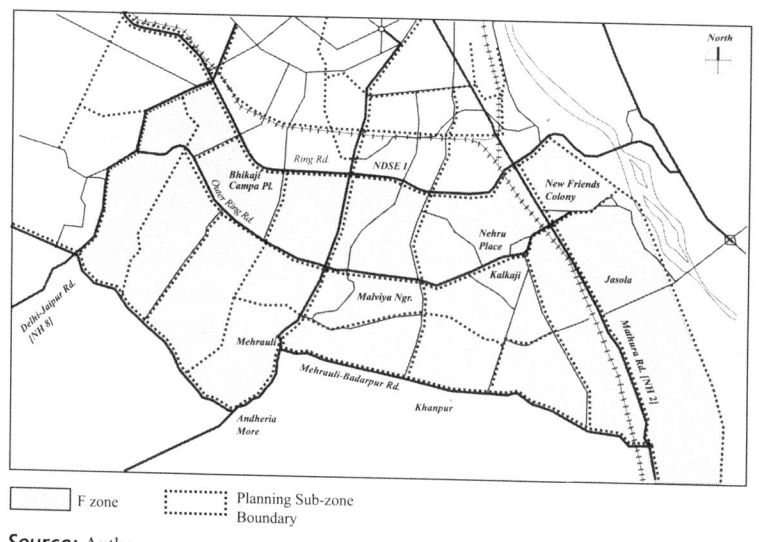

▢ F zone   ⦂⦂⦂ Planning Sub-zone
                    Boundary

Source: Author.

Figure 5.6:
Territories of the South Delhi PC within the F Zone

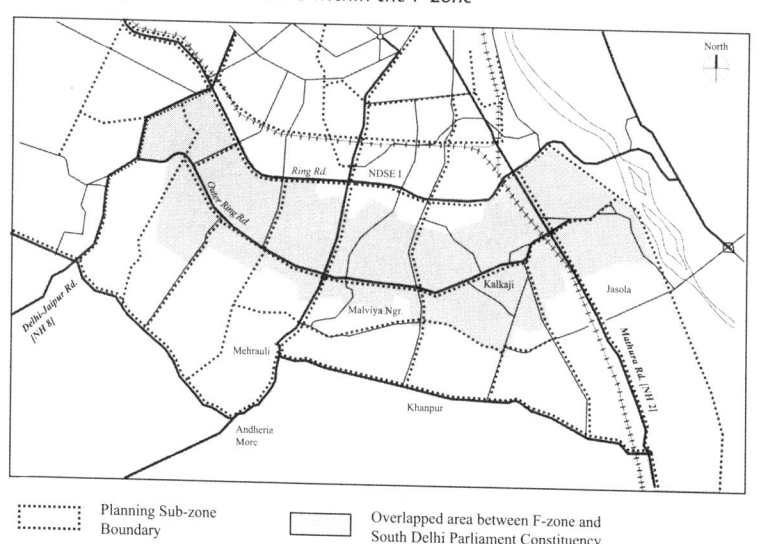

⦂⦂⦂ Planning Sub-zone
         Boundary

▢ Overlapped area between F-zone and
     South Delhi Parliament Constituency

Source: Author.

Note: South Delhi PC map is based on the jurisdiction before delimitation in 2006–2008.

Figure 5.7:
Overlay of ACs within South Delhi PC and F Zone

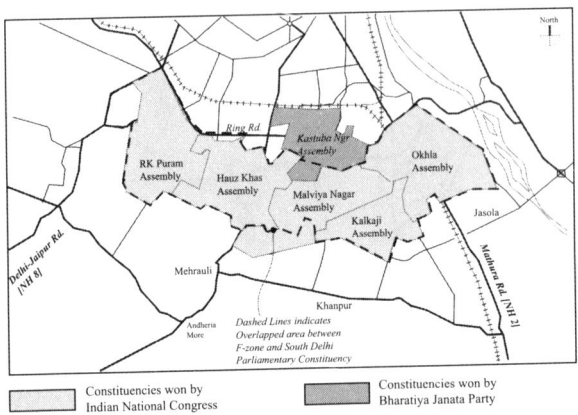

Source: Author.

Notes: 1. Shades of gray color indicate the winning political party in the Assembly election 2003.

2. ACs in this map are based on the jurisdiction before delimitation in 2006–2008. Two ACs, namely Malviya Nagar and Kasturba Nagar, fall partly within the South Delhi PC.

Figure 5.8:
Overlay of MCD Wards within South Delhi PC and F Zone

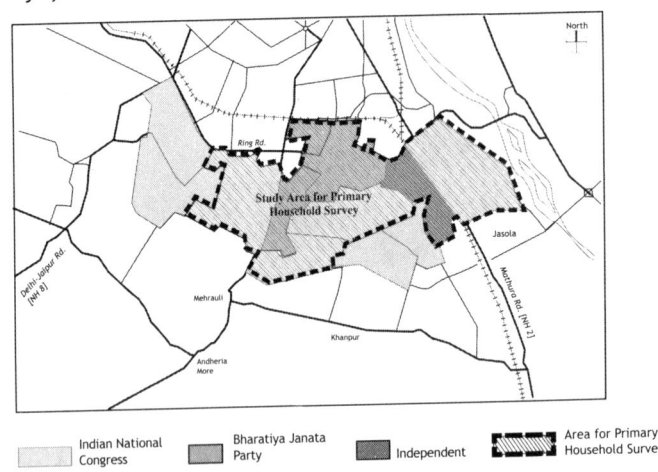

Source: Author.

Note: Shades of gray color indicate the winning political party in the MCD election 2002. MCD wards in this map are based on the jurisdiction before delimitation in 2006–2008.

Figure 5.9:
Colonies Included in the Primary Field Survey

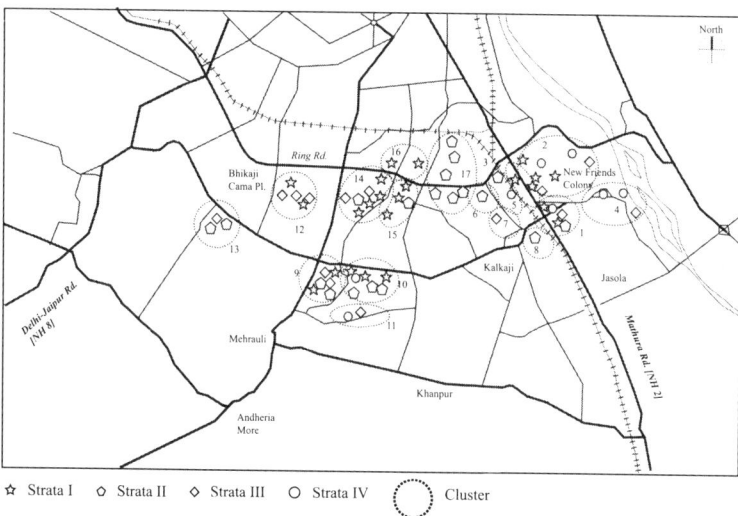

☆ Strata I   ○ Strata II   ◇ Strata III   ○ Strata IV   ⟨ ⟩ Cluster

Source: Author.

Notes: 1. To begin with, property tax categories (A to H) are mapped spatially and adjacencies of colonies with different tax categories are observed. Patterns of *adjacency* and *concentration* of different tax categories within a MW are identified and termed as "clusters."

2. Clusters are indicated in the above map by dotted boundary and each cluster is numbered.

3. Subsequently, the household survey is conducted in eight such clusters, four from the MW of Okhla and the rest from that of New Friends Colony.

4. Assembly Constituencies in this map and in the legend are based on the jurisdiction before delimitation in 2006–2008.

5. Clusters 1–8 are in Okhla AC, clusters 10 and 11 are in Malviya Nagar AC, cluster 9 is in both Malviya Nagar and Hauz Khas ACs, clusters 12–14 are in Hauz Khas AC, and clusters 15–17 are in Kasturba Nagar AC.

address specific questions, such as the delivery of water supply at the household level or per capita level, income-expenditure-savings of the households, and spending of political funds across various property tax strata of the colonies. The household and per capita, then, became the smallest component within the unit. For plotted colonies, individual plots and for flatted housing, individual flats were identified. In the plotted colonies, any one of the households in the chosen plot was selected, because each of them would have the similar locational condition. Units were chosen through stratified random sampling.

Figure 5.10:
Colonies Where the DJB Data Is Taken and the Statistical Analysis Is Conducted

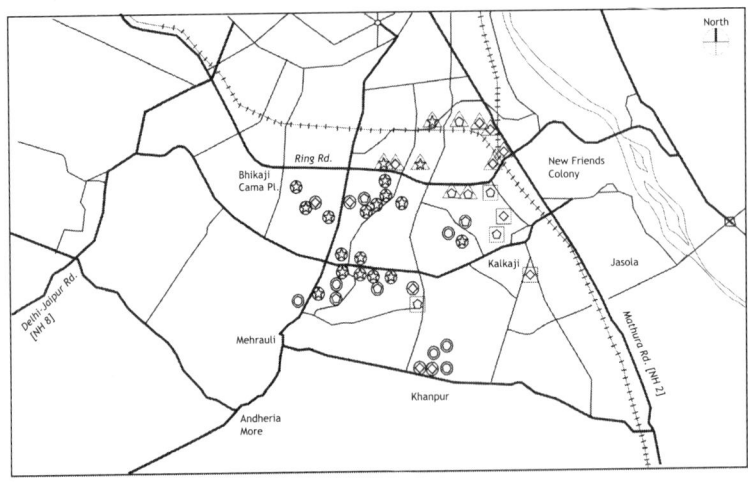

DJB Underground Reservoir (UGR)

	Greater Kailash I	Giri Nagar	Jal Sadan
Strata I	⊛	☆	★
Strata II	⬡	⬡	△
Strata III	⬨	◇	◇
Strata IV	◎	▢	△

Source: Author.

Note: Assembly Constituencies in this map and in the legend are based on the jurisdiction before delimitation in 2006–2008.

For quantitative analysis, statistical methods, such as the descriptive and visual analysis, the correlation, the factorial analysis of variance (ANOVA), and the factor analysis, are used. Five sets of data are considered for the quantitative analysis: the information on the socioeconomic strata based on survey data, the provider-end information on the delivery of water supply based on the DJB data, the user-end information on the delivery of water supply based on the survey data, the MLALAD funding through the DJB based on the DJB information, and the MLALAD funding through the MCD based on the MLA's declaration and the MCD information.

Table 5.3:
Details of Sample Size

Phases of the Sample Survey Conducted	*Sample Size (Plots/ Flats)*	*Total Number Covered (Plots/ Flats)*	*Assembly Covered*	*MW Covered*	*Total Number of Households as per Census 2001*[a]
Phase I	504	9,345	Okhla	Okhla, New Friends Colony	
Phase II	973	19,350	Malviya Nagar, Hauz Khas, Kasturba Nagar	Malviya Nagar, Greater Kailash1, Hauz Khas, Gulmohar Park, Sewa Nagar, Defence Colony	
Total	1,477	28,695			114,365

Source: Author.

Notes: Approximately 10 percent samples from each tax category from a particular cluster have been chosen and in both the phases, sample sizes are significant (>95 percent level of confidence) for representing the total number covered; even at the overall household level (N = 114,365 and n = 6,065). The total number of plots/flats covered in the sample (28,695) seems sufficient (>98 percent level of confidence) as per Yamane's (1967: 886) simplified formula to calculate sample sizes: $n = \dfrac{N}{1 + N(e)^2}$, where e = Level of precision and N = Population size (cited in Israel 2009[1992]).

[a] Census 2001 information is sourced from http://des.delhigovt.nic.in/Census2001/wards.htm (accessed March 10, 2008).

The first dataset on the socioeconomic strata was compiled on the basis of phase II of the primary household survey, conducted in 2007 over 43 colonies across three ACs, namely Kasturba Nagar (AC 4), Malviya Nagar (AC 8) and Hauz Khas (AC 9), and across seven planning subzones. Two of these planning zones were in the planning division D and the rest in division F. In this dataset, interrelationships have been established between the property tax strata and the socioeconomic status, indicated by the family-wise gross monthly income, monthly expenditure on education, electricity bill, transport and entertainment, and the average monthly savings. A strata-scale has also been formed in which strata I includes the property tax categories

A and B, strata II for categories C, strata III for E and F, and strata IV for G and H.

The second dataset included the supplier-end information on the delivery of water supply collated around 2007–2008 at the colony level.[7] Physical spaces covered in the data were served by two DJB service zones, South I and South II. Three major underground reservoirs (UGR), namely GK1, Giri Nagar, and Jal Sadan, located within these two service zones, supplied water to colonies. The operative framework for the collection of the colony-wise data from the DJB was comprised of the information on the resource allocation, the efficiency of the system, and effectiveness of the resource distribution. Relevant quantitative data were converted into population-based figures for statistical analysis. To begin with, the data on 84 colonies were gathered. However, data characteristics for several colonies were not found suitable for the population-based conversions to arrive at the factors of water supply. For example, information on the population or on the quantity of water supply was available for two or more colonies together, or population figures were unavailable for some colonies. Data have been analyzed for more than 60,000 direct domestic connections (61,262) in 48 ($N = 48$) colonies having more than six lakhs population (609,561). These DJB colonies have been spread over seven ACs and the planning divisions D and F (Figure 5.10).

The third dataset on the user-end information on water supply was gathered from the household survey during 2006–2007. Interview was conducted for 1,477 households across 63 colonies in 17 survey clusters. These colonies were spread over political jurisdictions of four ACs and eight MWs, four DJB service zones, 11 planning zones of which two were within the planning division D and the rest in the planning division F. Colonies included property tax zones from A to G covering all the four strata. The DJB piped supply has been the primary source of water for the households surveyed.

The fourth dataset was collated on the project-specific information available on the DJB website on the spending of MLALAD through DJB on water-related works. The DJB has the budget estimates under the "Special scheme for Grant-in-aid for Development of Sewerage and Water Supply" in ACs in which each MLA can suggest works for water supply and sewerage improvements to the extent of ₹ 50.00 lakh in their ACs. The MLALAD fund was spent through the DJB under two heads, the MLA fund and the MLA priority fund. Under the MLALAD fund, MLAs would allocate funds to the DJB and the DJB, then, decided on which project within the concerned MLA's AC the money was to be spent. Nevertheless, MLAs could always give their suggestions on the projects, and to consider political influence being completely oblivious to such spending would be naive. Under the MLA priority fund, the improvement works recommended by each MLA, would be taken up by the DJB. Data for a period of six years, spanning from 2004 to 2009, have been presented at the colony levels by taking work-wise locations for the DJB South II and III zones. Data were observed over 77 residential colonies within seven ACs, namely Dr Ambedkar Nagar (AC 34), Hauz Khas (AC 9), Jangpura (AC 5), Kalakji (AC 7), Kasturba Nagar (AC 4), Malviya Nagar (AC 8), Okhla (AC 6), three (3) first level UGR of DJB, namely GK1, Jal Sadan and Giri Nagar, and 14 planning zones of which four were from the planning division D, eight from the division F and one from the division J. For about 200 projects ($N = 197$), the total amount was allocated and for about half of those projects ($N = 105$), data on the total amount allocated per capita information were available.

The fifth dataset was prepared by spending the MLALAD fund through the MCD. The major share of the MLALAD, sometimes even more than 90 percent of the funds available, has usually been spent through the MCD. Hence, it was felt pertinent to observe the overall pattern of spending of this particular political fund with relation to the socioeconomic strata for any generic trend. For this purpose, the status report of the MLA fund by the MCD, available at the agency website, was used as the data source for the project-specific information on the spending of the MLALAD through MCD (MCD 2010). Data were observed for over 700 projects ($N = 710$) within a time period from 2004–2005 to 2007–2008 over 78 colonies spread across four (4) ACs, namely Hauz Khas (AC 9), Malviya Nagar (AC 8), Kalakji (AC 7), and Kalakji (AC 6), 16 wards and 11 planning zones in the planning division F. Data were presented at the colony levels in this research by taking work-wise locations within the study area for the MCD South and Central zones.

I shall now discuss the relationships between involved socioeconomic spaces/strata and the notion of property.

Notes

1. MPD, albeit at a simplistic operative level, identifies infrastructure as a threefold operation: physical infrastructure, social infrastructure and transport, and includes water supply as part of physical infrastructure.

2. "Informal settlements," as popularly termed, have been "built outside administrative procedures," namely the Master Plan (The Urban Institute 2007: 36). The Urban Institute (2007) in its report, "A New Land Title Registration System for Delhi: Recommendations," identifies informal settlements, and its land tenure have certain characteristics: "no title for the land parcel requested or obtained from the titling authority (at the titling level) or no building permit requested or obtained from the urban planning authority (at the construction level), and thus, facing a risk of demolition." This report, like others (e.g., Dhar Chakrabarti 2001), identifies certain settlement types within the broad classification of *informal settlements* of Delhi, namely the squatters/JJ clusters, resettlement, and relocation colonies, unauthorized colonies, rural and urban villages, and gives detailed discussions on the tenure of these informal settlement types (ibid.: 36–38).

3. The MCD has been trifurcated in 2012 and the newly formed corporations are: South Delhi Municipal Corporation, North Delhi Municipal Corporation, and East Delhi Municipal Corporation. This has happened much later than the time of this empirical research. The reference of "South Delhi" is rather colloquial and to some extent, geographical and must not be mixed up with the South Delhi Corporation.

4. At the time of the empirical research, before the delimitation in 2006–2008, three Parliamentary Constituencies (PCs), six ACs and 11 Municipal Wards (MWs) were completely or partially located in the F zone. Along with that, about 11 DJB water supply subzones are also situated within the same zone.

5. Elsewhere, the relationship between the wealthy and the urban poor, or the mill-owners and the workers, or the mediators, such as the non-charitable organizations or the trade unions, and their beneficiaries, such as the urban poor or the workers, are referred to as the "patron–client relationships" (Chakrabarty 1989, Chatterjee 2004: 132).

6. Apart from the household survey, the Resident Welfare Association (RWA) survey and the reconnaissance survey were also conducted. The RWA survey, conceived to get some specific information to supplement the household survey of that particular colony, unfortunately did not get sufficient responses and has not been considered for further analysis. Most of the office-bearers were unavailable for comments for various reasons. Only RWAs of 11 colonies out of 63 colonies agreed for the interview and that too without providing answers for several important issues. The reconnaissance survey primarily recorded the urban form along important street edges to indicate

qualitative aspects of spaces, what Lynch (1960) terms as the "imageabilty," as well as the condition of the overall maintenance of the colony through photographs, and in few occasions, important locations were marked with the help of annotated maps.

7. Colonies enlisted in the South I and II zones of the DJB as per the DJB website in 2007 were considered.

6

Socioeconomic Spaces and Property Rights

Here, the central question is: What is the correspondence between legal property rights and socioeconomic spaces? My argument is twofold, regarding the multiple gradations of socioeconomic spaces and the use of the property tax category as a possible surrogate indicator of the socioeconomic spaces in the city of Delhi. In the absence of any direct measure of socioeconomic spaces, property tax strata, based on the tax categories set by the MCD, can be used as the appropriate surrogate indicator to classify the socioeconomic spaces of the residential setting of Delhi.

The argument is arranged in three broad segments: The notion of property in forming the intended city and the real city, the elaboration of the property tax system as a legal framework in Delhi, and the empirical relationship between the property tax strata and the socioeconomic spaces in Delhi. The hypothesis emerging out of the empirical analysis is: Higher property tax strata represent spaces occupied by higher socioeconomic strata.

Property and Community

The basic framework of the rights of the democratic state is the notion of "égaliberte", equality and liberty, as elaborated earlier (Balibar 1994, Chatterjee 2004). Both equality and liberty are included in the Preamble of the Constitution of India as well. If any of these rights is hampered, the condition of justice would be disturbed. Balibar (1994, 2005: 21) identifies that, besides other normative qualifications or disqualifications, the rights of citizenship and residence (i.e., the property rights) are central to the notions of exclusion and inclusion in the society.

Property provides an individual the status in the form of citizenship, which allows one to be part of a *community* (Balibar 2005). *Community* is the collective totality and any contradiction of rights is to be addressed to the whole of it (Balibar 1994, 2005, Chatterjee 2004). Habermas (1995: 127–128), the noted German sociologist, too, includes property rights alongside other liberal rights of the "liberty of belief and conscience," "the protection of life," and "personal liberty." Having deliberated further on Balibar's observations on the *property* and the *community*, Partha Chatterjee (2004: 74–75) introduces in his book, *The Politics of the Governed*, the dual existence of the "civil society" and the "political society" of citizens and populations, respectively—the former occupies the realm of theory and the latter, of policy.[1] The legal (or the legitimate) bindings and the para-legal (or political) arrangements are referred to as the respective conditions of the civil and the political society.

The civil society's association with the citizenship, community, and the political society seems changing and besides political and civil liberties, the civil society tends to include "private property rights and markets" as well (Khilnani 2001: 11). The political society accommodates "an entire set of para-legal arrangements that can grow in order to deliver civic services and welfare benefits to population groups whose very inhabitation or livelihood lies on the other side of legality" (Chatterjee 2004: 56). Although para-legal situations are not often legal in the eye of the policy makers or planners, yet they manage to get "recognition as a population group" (Chatterjee 2004: 57). Such thoughts explain the services of water, sanitation, and electricity provided by civic authorities to the squatters. In particular, within the hegemony of urban planning in Delhi, these coexistent realities of the para-legal, working outside/alongside the defined domain of the legal, and the legal itself, create what I refer in this work as *the real city*. The city according to the planning document is conceived as *the intended city*.

Even within the legal parts of the city, I argue later in this book, multiple gradations of equity and justice are formed. Real parts of the city that contain the so-called legal existences in the form and notion of planned colonies or legal property rights or the citizenship or the civil society, all having equal legal rights in accessing state-supplied resources, receive differential distribution and delivery of urban services of water supply from the state. That, indeed, opens up the discourses of multiple shades of urbanity in place of prevailing *two-city* notions.

Two-city Notions on Delhi

The prevailing notions on the city of Delhi are based on similar binary oppositions starting with the planned and the unplanned city. Planning, here, is also a legal tool, thereby imparting the legal justification of the planned city, whereas the unplanned city is what has grown outside of and in spite of the planning provisions. Dominant views of the government authorities, such as the Delhi Development Authority (DDA) and other administrative authorities, on these dual conditions of living are broadly Malthusian in its apprehension: More people means deterioration of resources. In fact, most of the authorities observe the *unplanned* living conditions as the way of living by the poor, who are also considered the violators of the Master Plan by encroaching or polluting the environment (Dhar Chakrabarti 2001, Vedeld and Sriddham 2002).

A rich body of work argues on the dichotomous status of the city of Delhi from the other end of urban existences. The continuous plight of the urban poor in Delhi is viewed as the result of the lack of implementation of the MPD 1962 and related bad governance. The shortfall of the housing provision for the LIGs by the state is held accountable for the condition of the poor today, and the attitude of urban planning as well as of the government agencies toward them and their habitat are also highly criticized (Baviskar 2002, Bhan 2009, Dupont 2004, 2008, Ghertner 2008, Kumar 2006, Roy 2000, Tiwari 2003, Sharan 2006, Sundaram 2010, Verma 2000). An interesting observation on this issue is that the subsistence of the urban poor in Delhi is formed out of the "poverty or need", or the "legal citizenship", or "his contribution to and work within the city" (Bhan 2009: 133). This is what Chatterjee (2004) describes as the "para-legal" status of urban living. Thus, "the state of exception", Agamben (2003[2005]) concludes, comes into play out of the condition of necessities.

Therefore, the real city on ground today is a heterotopic coexistence of what was intended by the Master Plan and built accordingly, the legal, and what was not, the illegal/para-legal. In Delhi, both these conditions are treated differently in accessing civic amenities and facilities.

Property Tax Strata in Delhi

Now, it may be useful to consider the property tax as an indicator of the legal notion of property in Delhi. The property tax is understood as "a unique mechanism for local revenue generation" for property that is "visible, immobile and a clear indication of one form of wealth" (McCluskey 1999, in Mathur et al. 2009: 13). My concern, here, is limited within the so-called *legal* parts of the city where I demonstrate the correlation between the multiple socioeconomic strata and the property tax strata of residential colonies.

Local urban governance in Delhi is managed by three agencies, the MCD, the NDMC, and the Delhi Cantonment Board (DCB) over respective territorial jurisdictions (Figure 6.1). Among these three

Figure 6.1:
MCD Zonal Set-up

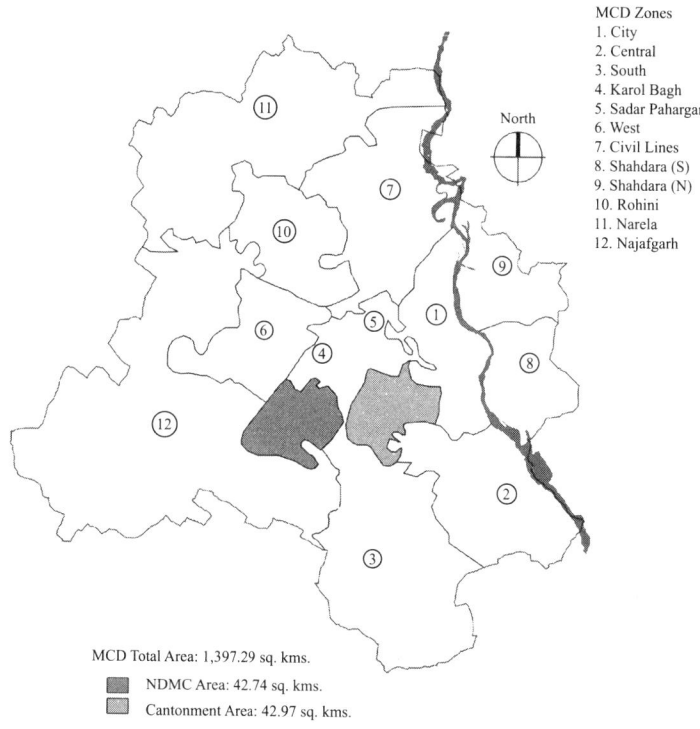

MCD Zones
1. City
2. Central
3. South
4. Karol Bagh
5. Sadar Paharganj
6. West
7. Civil Lines
8. Shahdara (S)
9. Shahdara (N)
10. Rohini
11. Narela
12. Najafgarh

MCD Total Area: 1,397.29 sq. kms.

█ NDMC Area: 42.74 sq. kms.
▨ Cantonment Area: 42.97 sq. kms.

Source: Adapted from MCD (2005).
Note: In 2012, MCD was trifurcated into North DMC, South DMC, and East DMC.

agencies, the MCD has the largest jurisdictional area covering more than 94 percent of Delhi (MCD 2005). The property tax strata considered in this work hinges upon the concepts and guidelines of the MCD system. The unit area method of property tax, implemented by the MCD, from April 1, 2004, has been "designed to be citizen-friendly, transparent, easy to calculate, promote honesty in the citizen–civic body interface" (MCD 2005: 36–39). Property tax, levied on all lands and buildings within the municipal area of Delhi, is also the major source of income of the MCD; for example, about 64 percent of the income in the budget of the MCD for the year 2005–2006 was estimated from the municipal taxes (ibid.). According to the property tax method, the city of Delhi is grouped into categories of colonies with specific unit area values (UAV). The UAV is computed on the basis of the category of colonies, the type of structure, the age of construction, the use of the property, and the occupancy factor, and for each of these parameters, a multiplying factor is given.[2] JNNURM (Jawaharlal Nehru National Urban Renewal Mission), too, in its guidelines for the urban tax reform discusses the unit area method of property tax as a possible system (ca. 2005).

A study, sponsored by the 13th Finance Commission of the Government of India, observes four types of property tax systems in the country (Bahl and Linn 1992, Lall and Deichmann 2006):

> One is the annual rateable value (ARV) of lands and building, second is a variant of the ARV where the determination of ARV is with reference to location, type of construction, age of building, and the nature of use to which a property is put; third is capital valuation, recently brought in use in Karnataka, and the fourth is direct computation of property tax by using a tax rate per unit of carpet area. (Mathur et al. 2009: vii)

In Delhi, UAV, a variant of the annual ratable value has been adopted as the property taxation method and likewise, in several cities and states in India, simplified systems are in use for the property tax evaluation.[3] Nevertheless, the most important component of the property taxation is, indeed, the grouping of the localities and colonies in a category. The categorization of colonies in Delhi was made on the basis of 10 parameters: capital value of land, prevailing rental values, age of the colony, physical infrastructure, social infrastructure, proximity to commercial centers, main road on which the colony is located, type of settlement, economic status of occupants and location (Naresh and Halen 2006).[4] JNNURM Primer (ca. 2005) also indicates similar principles for such categorization. Each parameter has been graded either A, B, or C; the grade scales were then converted to a point scale (A = 10, B = 6, C = 2) and, finally, a classification matrix was prepared for the categorization of colonies/areas (Naresh and Halen 2006).[5]

Colonies were then "geo-coded" and plotted on the digital map and, thus, the MCD has grouped the colonies of Delhi into eight categories of localities from A to H, A being the highest and H, the lowest. Rural villages are placed in the lowest category of H and the categories of urban village are *location-specific* by being placed two slots below the category of the highest neighboring colonies (MCD 2005: 69). The UAV of the property tax is stable for three years and is to be revised every three years based on the recommendations of a Municipal Valuation Committee. As per the report of the first Municipal Valuation Committee in 2004, the UAV are recommended based on the model in which the UAV of the category D as ₹320 is the base value, 25 percent is added for the UAV for higher categories, 15 percent is deducted for lower categories, and values rounded off to the nearest multiple of 10.[6] Thus, there is a hint of stratification on either side of the D categories.

The first Municipal Valuation Committee (MVC-I), having incorporated public objections and suggestions, recategorized 1986 colonies (Municipal Valuation Committee-III 2010). However, the Hardship and Anomalies Committee (HAC) further reviewed that categorization and finalized the category-wise list of 2,025 colonies, based on which the property tax is presently being collected (Municipal Valuation Committee-III 2010).

Property Tax Strata as the Socioeconomic Strata

Interestingly, an order by the Honorable Justice Sanjiv Khanna at the High Court of Delhi in his order (dated January 14, 2010) reaffirmed the importance of "prime" location, "prestige", property values as the "premium" for categorizing a colony, all of which were also the indicators of the socioeconomic status of a residential colony.[7] The circle rates for the valuation of land for the residential use in Delhi, stipulated by the revenue department of the state government, specify eight categories of A to H, which, in fact, correspond to the categories for the minimum rate for construction as well (Table 6.1). The property tax categories, the circle rates, and the construction rates are worked around the same slabs of A to H

Table 6.1:
Comparison among Categories of Locality

Category of Locality	Circle Rate for Valuation of Land for Residential Use[a] (₹ per sq. meter)	Circle Rate for Valuation of Land for Residential Use[b] (₹per sq. meter)	Minimum Rate for Construction for Residential Use[b] (₹ per sq. meter)	UAV[c] (₹ per sq. meter)
A	86,000.00	43,000.00	14,960.00	630
B	68,200.00	34,100.00	11,870.00	500
C	54,600.00	27,300.00	9,500.00	400
D	43,600.00	21,800.00	7,600.00	320
E	36,800.00	18,400.00	6,410.00	270
F	32,200.00	16,100.00	5,600.00	230
G	27,400.00	13,700.00	4,750.00	200
H	13,800.00	6,900.00	2,370.00	100

Source: Compiled by the author.
Notes: Circle rates were introduced from 2007.
[a]Times News Network (2011a).
[b]Minimum rates (circle rates) for valuation of land and properties for the purposes of registration under the Registration Act, 1908, in Delhi. Available at http://districts.delhigovt.nic.in/circlerate&cases.PDF (accessed March 17, 2003).
[c]As per recommendations of the Municipal Valuation Committee-I, 2004.[8]

and are the points of coincidence of the property valuation and the property taxation fixed by the state and the municipal governments, respectively.

The economic stratification of living, for example, LIG, MIG, and HIG, has been one of the popular parameters of spatial and locational distributions of different income groups in the city. Delhi is no exception to that. The housing delivery of the DDA has followed similar strategies as well. But the property tax categories of residential colonies by the MCD make an alternative stratification.

Since, there are realistic possibilities of similar socio-income groups residing in immediately higher or lower categories, two adjacent property tax categories are grouped into one stratum. Strata I includes property tax categories A and B, strata II, categories C and D, strata III, categories E and F, and strata IV, categories G and H.

I argue that the income of the residents has a positive correlation with the property tax strata of a colony. Furthermore, extending the argument beyond the relationship between income and the strata, one may say that the socioeconomic status of the residents of a colony has certain relations with the strata of that colony. The socioeconomic status of the residents of a colony is indicated by the combined factor of the gross family income per month, monthly expenditure on certain accounts, such as education, electricity, transport (or petrol) and entertainment, and the average monthly savings.

A sample survey over 946 households across 43 colonies, spread over three ACs, namely Kasturba Nagar (AC 4), Malviya Nagar (AC 8), and Hauz Khas (AC 9) and six MWs, show that a wide gap exists in the average monthly family income across the strata. Average monthly income of strata I (₹86,405.84) is about seven times of that of strata IV (₹12,432.43) (Table 6.2, Figure 6.2). A common sense observation would tell that the average income figures found out in the survey especially in higher strata, such as strata I and II, seem less. One reason would be people's reluctance in disclosing their income. Nevertheless, the average monthly income is clearly higher in higher strata. Average

Table 6.2:
Strata-wise Comparison of Income, Expenditure, and Savings

| Strata | Average Gross Family Income | Average Expenditure on | | | | Average Savings |
| | | Educa-tion | Electric-ity Bill | Trans-port/ Petrol | Entertainment | |
	(₹ per Month)	*(₹ per Month)*	*(₹ per Month)*	*(₹ per Month)*	*(₹ per Month)*	*(₹ per Month)*
I	86,405.84	9,101.23	5,250.42	5,434.26	4,544.26	13,352.59
II	61,265.92	4,383.99	2,289.96	2,540.99	2,477.01	7,885.92
III	35,013.76	2,155.08	1,349.65	1,754.15	1,366.87	2,842.07
IV	12,432.43	852.21	625.54	1,101.52	852.34	1,522.73
Grand Mean	63,325.05	5,597.24	3,154.74	3,487.24	3,062.22	8,429.12
N (No. of households)	946	825	924	887	803	577

Source: Author.
Note: Data is based on the survey responses. But one may see that absolute figures seem low especially in strata I. However, overall stratification is clear even with such low figures.

Figure 6.2:
Strata-wise Mean Family Income, Expenditure, and Savings (₹)

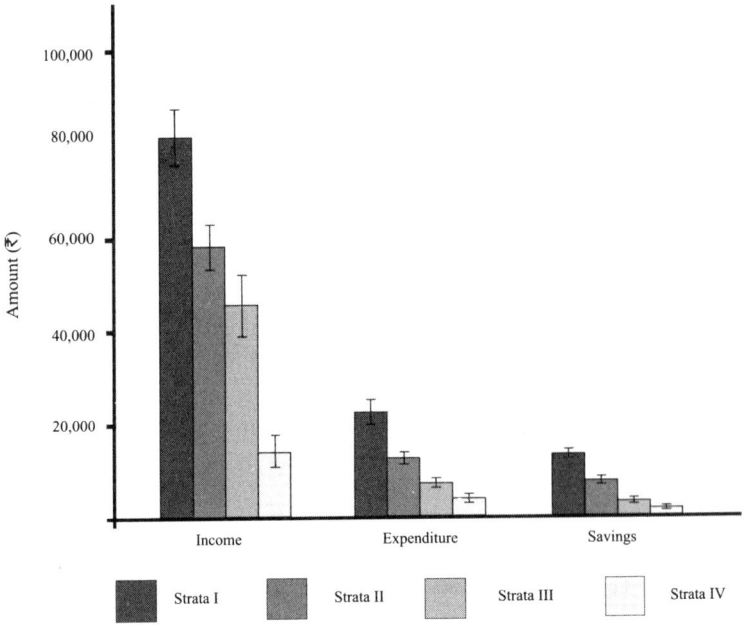

Source: Author.
Note: Expenditure includes expenses on education, electricity, transport, and entertainment only. Bar chart above clearly shows higher income, expenditure, and savings in higher strata. Also, 95 percent confidence intervals have no overlaps and grow toward higher strata thereby indicating clear-cut separations between each strata.

monthly expenditure on some of the important heads, such as education, electricity, transport, and entertainment, which are about 25 percent of the overall average income, also indicate higher expenditure in higher strata (Table 6.2, Figure 6.2). Average monthly savings are higher in higher strata too. In fact, the average monthly savings in strata I are almost nine times of that of strata IV (Table 6.2, Figure 6.2). ANOVAs with each of these variables also show the significant effect of the strata on the income [$F(3,942) = 118.352, p < 0.0001$], the expenditure on all four heads [$F(3,713) = 85.891, p < 0.0001$], and savings [$F(3,573) = 104.431, p < 0.0001$].

To find out the composite effects of the variables, factor analysis has been adopted to evolve a single socioeconomic factor involving gross family income, expenditure (on education, electricity, transport/petrol, and entertainment), and the savings.[9] Figure 6.2, showing the strata-wise mean value of the socioeconomic factor, is a very clear indication of the higher strata having higher factor (Figure 6.3). ANOVA results, too, show that the strata has highly significant effect on the socioeconomic factor [$F(3,481) = 72.116$, $p < 0.0001$]. Tamhane's post hoc tests also show highly significant mean difference between each strata ($p < 0.001$) and that higher strata has higher mean.

All these confirm that the strata are indicators of socioeconomic groups and can be used as the surrogate indicator for the socioeconomic status of a colony; then, the respective colonies can be considered as the corresponding socioeconomic spaces.

Figure 6.3:
Strata-wise Mean Values of the Socioeconomic Factor

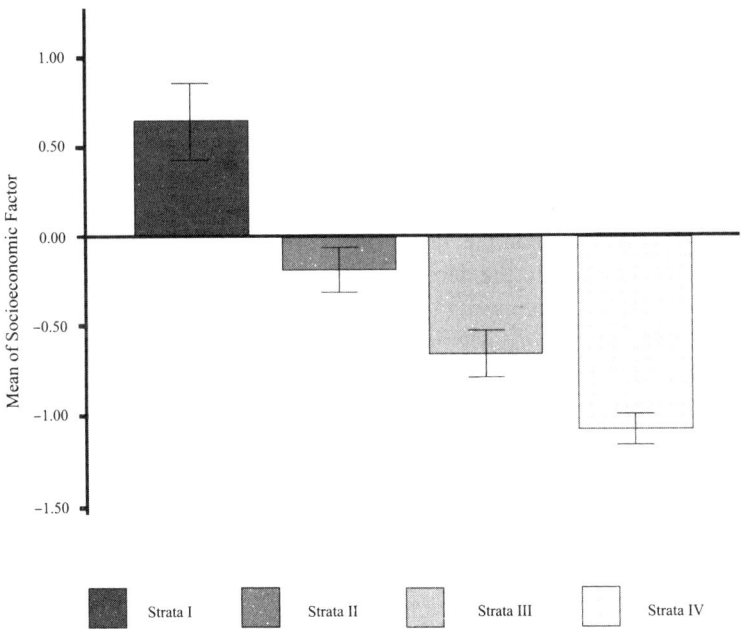

Source: Author.
Note: Ninety-nine percent confidence intervals have no overlaps and grow toward higher strata.

Observations

When the property tax strata represent the socioeconomic strata of people and the spaces (colonies) they occupy, it may be used as a qualitative parameter in a statistical analysis. Property tax strata would, then, be a representative indicator of spaces and people in place of the conventional data based on income and expenditure. As a result, it would be possible to compare spaces across the city. Also, once the tax strata are mapped across the city space, the observation of its disposition would correspond to the coexistence of multiple groups of people from different socioeconomic strata. Now, the question is: Despite the physical proximity to each other, do these diverse strata have the same level/degree of access to the resources and services? If it is not so, the notion of locational proximity requires a close look.

Interestingly, the strata, perhaps, can be used like the concept of the "income charge" that Harvey (1975[1973]: 53–54) introduces for taking into account the combination of earning (more or less), benefits (positive or negative), resource availability (more or less), and price of resources (higher or lower). The property tax strata, being the surrogate indicator of the socioeconomic spaces, seem to consider the aspect of *earning*. Decisions on *pricing* (or cross-subsidy) of resources may also be related to the strata, for example, the lower strata to pay less than the higher ones for per unit of resources. Such notions may lead to the decisions on *charging pricing premiums for higher strata of the society*. These conceptions, in fact, may guide the kind of *benefits* people enjoy. The benefit question may be addressed in three ways: Do all the strata have similar availability of (or accessibility to) resources and services? If they do not, it relates to the next question: Which strata do have less supply of resources and services and which ones have more? If lower strata of society are found to have less opportunity to access the resources and services, they would have less *capability* of mitigating such a shortfall. The condition of inequity and injustice is, thus, created. I shall discuss some of these questions with a specific reference to the basic urban services of water supply in Delhi.

Notes

1. Partha Chatterjee's work, *The Politics of the Governed* (2004), a highly significant narrative of the contemporary cities, looks at these two societal formations as somewhat clear-cut binary opposites in which nuanced observations are made particularly with respect to the political society formations.

2. The Section 116E of the Delhi Municipal Corporation Act 1957 (amended in 2003) specifies the following:

   > The annual value of any covered space of building in any ward shall be the amounts arrived at by multiplying the total area of such covered space of building by the final base UAV of such covered space and the relevant factors as referred to clause (b) of sub-section (2) of Section 116A. (MCD 2005: 9)

   Following are the steps in calculating the property tax (ibid.: 36–39):

   a. Measure the covered area of the property.
   b. Take the UAV of the locality/category notified by the corporation.
   c. Calculate the annual value [= UAV × covered area × multiplicative factors (OF, AF, SF, UF)], where, OF: occupancy factor; AF: age factor; SF: structure factor; UF: use factor.
   d. Calculate the property tax [= (annual value × rate of tax) – rebates/concessions as applicable].

   For residential properties, occupancy factor (OF) = 1 and rates of tax = 10 percent for categories A–E and 6 percent for F–H (MCD 2005: 69).

3. In Patna, three grades are made as per the street size; in Ahmedabad, the city is classified into four broad categories based on land value; in Hyderabad, the average rental value for each locality for each type of use is prescribed, whereas Karnataka is to use a capital-value-based system (JNNURM ca. 2005, Mathur et al. 2009).

4. The interim report of the Municipal Valuation Committee-III (2010) has recently recommended the inclusion of the 11th parameter, that is, the location within 1 km of Metro line.

5. The following point-range for categorization of colonies/areas were estimated (Naresh and Halen 2006):

Grades	80 and above	70–79	60–69	50–59	40–49	30–39	Less than 30	Rural villages
Categories of colonies	A	B	C	D	E	F	G	H

6. The UAV of the property tax is stable for three years and is to be revised every three years based on the recommendations of a Municipal Valuation Committee as per the amended provisions of the act.

7. Based on a recommendation by the Hardship and Anomaly Committee of recategorization of Siri Fort Bungalows from category A to category C, Siri Fort Road Residents Welfare Association and Mr C.P. Sabharwal, the President of the Association, filed a writ petition. The petitioners cited examples of colonies like Raghubir Nagar, Rajindra Park, Pant Nagar, etc., for which recommendations of recategorization had been accepted by the MCD. However, the MCD, the respondent in this case, neither recommended nor approved the recategorization of Siri Fort Bungalows from category A to category C. Hon'ble Justice of the High Court of Delhi, Mr Sanjiv Khanna in his order (WPC No. 4905/2007), dated January 14, 2010, stated the following concluding remarks:

> It is a matter of common knowledge that Siri Fort Bungalows are located at a prime location and is a prestigious residential colony in South Delhi, which commands premium. The flats/houses located in the said area are one of the most expensive houses in Delhi. I do not find any merit in the present writ petition and the same is dismissed. (The High Court of Delhi 2010)

8. As per the recommendations of the Municipal Valuation Committee-I (2004), GNCTD. Portions of the report are mentioned in Dhawan (2004) and also in Municipal Valuation Committee-III (2010).

9. All inter-variable correlations are significant at 0.01 level ($p < 0.00001$). Correlation coefficients are not more than 0.9, thereby negating a possibility of multicollinearity. Also, the determinant (= 0.129) is greater than 0.00001. KMO statistic, which normally varies between 0 and 1, of a value (= 0.799) is good and suggests that factor analysis is appropriate for these data. Here, the significant test tells us that the R matrix is not an identity matrix. There are some relationships between the variables that can be included in the analysis. For this data, Bartlett test is highly significant ($p < 0.0001$), indicating the appropriateness of the factor analysis.

7

Water Supply and Socioeconomic Spaces

Now, the question is: How does the delivery of the urban water supply in Delhi correlate with city's socioeconomic spaces and its legal property rights? The research question reflects upon the central empirical component of this work and I shall make an attempt here to study whether any selectivity occurs in the pattern of delivery (or distribution) of urban resources and services across different socioeconomic spaces in the city.

Since the 1990s, in many sectors, such as health, power, banking, telecom, housing, etc., the welfare state policy of India has made room for the neoliberal economy and related policy decisions. This transition, however, continues to witness the simultaneous presence of local and global private enterprises. At the same time, the government organizations are also seen as parallel operators or as collaborators within the PPP model. Of late, in the field of urban planning, too, the City Development Plan (CDP) under the JNNURM identifies the PPP as a possible future model of urban delivery.

Delhi is one such site of all these *privatization* moves, yet water supply till date is provided by the state. Water, the most essential urban service for human life, is also a scarce resource in Delhi and, therefore, the concerns for equity in the delivery and distribution across socioeconomic spaces would be an overriding phenomenon. The delivery of water supply in Delhi, I try to argue here, significantly varies even across the so-called legal colonies (socioeconomic spaces) and such variations correspond to the socioeconomic strata represented by the property tax strata of the colonies. Planned colonies, being *legal* entities, are also included here, which, otherwise, are supposed to get similar quantity of supply as per the planning policy.

The hypothesis is: Same supply norms and legal property rights do not necessarily ensure equal water supply, and the real delivery varies across socioeconomic spaces within planned residential areas in a way that the higher strata of socioeconomic spaces (i.e., the colonies) get higher supply of water.

Access to Water Supply, Notions of Equity, and Social Justice

Conditions and degrees of accessibility to the piped water supply in the city (or the lack of it) remind us of the just, intended, and real distributive conditions, introduced earlier. The just and the intended conditions, following the ideological position and the policy framework, respectively, should match with each other in an ideal, rather utopian, situation and similarly, the real distribution should also match the intended policy (Figure 1.2). Selectivity arises when the policy does not correspond to the ideology, and the delivery to the policy. In turn, the *politics of distribution* is formed.

In case of water, a simplified understanding would tell us that cities in different climatic regions, or from completely different cultural habits, or with the availability/scarcity of raw sources of water, may have different levels of consumption and supply. If we can assume similar consumption of water for physiological, health, and hygiene needs of humans within the given climatic and cultural condition of a particular city, the quantity of water supply per capita in the just distributive condition should be unchanged across all social and income groups in that city. In such a utopian setting, just and intended distributions would be two sides of the same coin. But, realities are full of divergent practices of different sociopolitical contexts across various times and spaces (historically and geographically), where intentions have not been to ensure the just distribution across the society.

In feudal and cast-dominated societies in India, historically, the usage of water had been one of the major elements of social stratifications, traces of which might be seen in several rural settings of the country even today. Overall distribution of use zones along the river, for example, the temple at the upstream, bathing and other activities at the middle, and the *dhobi ghat* (the washing spaces) and the cremation ground at the downstream as in Ujjain, as well as the socio-spatial segregation of the *ghats* (the bathing spaces) along the river edges can be observed in many traditional Indian cities.

Intentions of similar historical–geographical practices can also be found in contemporary water discourses that narrate the interplay of spatial, social, cultural, ecological, political, technological, and governmental issues leading to variable conditions and degrees of accessibility to water (or the lack of it) (Bakker 2003a, 2005, Gandy 2004, Kaika 2005, Keil 2005, Kooy and Bakker 2008, Swyngedouw 1999, 2004, 2006). Most of these observations, indeed, underline the point that "[w]ater is a brutal delineator of social power which has at various times worked to either foster greater urban cohesion or generate new forms of political conflict" (Gandy 2004: 363).

Some of these works are, for example, the case studies of the "development and differentiation" of the urban water supply in colonial and postcolonial Jakarta (Kooy and Bakker 2008) and in contemporary Israel and Palestine (Phillips et al. 2007), historical–geographical analysis of the socio-environmental implications of the water management in Chile under the military regimes and the democratic coalitions (Budds 2004), the historical analysis of colonial discourses of the "contaminated city" and public health debates and practices in Mumbai (McFarlene 2008), the historiography of the ecology of urban water in Athens and London (Kaika 2005), the generic discussion on urban water systems in the post-industrial cities across the world (Gandy 2004), or the discourses on larger issues of urban water supply through the analysis of the specific case of water politics in Ecuador (Swyngedouw 2004). These pioneering works are, often, concerned with the analysis at larger scale-levels encompassing the systems and networks of water from the standpoint of urban geography, and are thematically constructed on the framework of "political ecology" that elucidates multiple interrelated practices of political, ecological, socioeconomic, cultural, and technological forces implicating each other (Escobar 1999, Swyngedouw 2009).[1] In some of these works, traces of Foucauldian notions of bio-politics and governmentality can be found alongside historical–geographical materialism of the Marxist thinking (Escobar 1999, Gandy 2004, Kooy and Bakker 2008, Swyngedouw 2009). At the same time, there are also attempts to bring the Gramscian framework of the Marxist understanding and the Foucauldian notions together to analyze political ecology oriented discourses on water (Ekers and Loftus 2008). Perhaps, the geographer and urbanist Mathew Gandy's notion of the "bacteriological city" may explain the political ecology idea a bit clearer, especially with respect to water:

(…) a new socio-spatial arrangement that could simultaneously ensure a degree of social cohesion at the same time as protecting the political and economic functions of the modern city. Water played a pivotal role in this reconstruction of urban space to produce what we would recognize as an archetypal modern city with its closely choreographed intersection between technology, space and society. (Gandy 2004: 365–366)

If the technology of the piped water supply is similar across the city, alike in contemporary Delhi, space and society would be crucial for any selectivity in the distribution and access. All these make water supply an interface of socioeconomic and political forces in cities (Ekers and Loftus 2008).

Most of the discussions before are based on ideological conditions that do not necessarily conform to the concerns of the just access of water or its just distribution. However, the equitable distribution has been the central ideological scaffolding of the adoption of planning as a policy in the post-independence India and its subsequent application in Delhi. The explication of water supply in Delhi is to follow the issue of equitable distribution by the state and the accessibility to the supplied water by the people at the delivery ends. The broader question arises: What are the parameters of such a just distribution?

While interrogating the relevance of "social equity" considerations in the privatization process in water supply and sanitation services around the world, Isabelle Fauconnier (1999), an urban planner and economist, defines three parameters specific to the notion of equity in water services: Physical access to safe drinking water, economic access or affordability, and access to planning and decision-making for the services. While the first two parameters refer to the aspects of accessibility and affordability, respectively, the third one may be related to the planning and political processes. In general, concepts of equity may carry four different meanings (Fauconnier 1999, Musgrave and Musgrave 1984):

- The *vertical/distributional* equity based on the "ability to pay" or affordability principles;
- The *horizontal equity* related to the "benefit" principle that suggests the same price for the same amount of benefit for any good or service to different individuals across all groups;
- The *geographical equity*, a corollary of distributional equity in some cases, refers to the equitable or even distribution of services across different geographic locations (e.g., urban versus rural and center versus periphery); and

- The *intergenerational equity*, a concept useful for the evaluation of environmental impact of resource consumption (e.g., water), indicates that the consumption by present generations is not at the cost of consumption by future generations.

The concepts of vertical/distributional and horizontal equity can be understood by studying the planning considerations, that is, the policy of allocation of water at various scale-levels, as well as the implementation conditions, that is, the governmentality of the real delivery across diverse socioeconomic spaces. In effect, the phenomenon of socioeconomic space would be a useful variable accounting for both the *vertical* (the economic) and the *horizontal* (spatial) parameters of equity. These distributional concerns are to be situated now within the framework and the context of this work.

An Overview of Water Supply in Delhi

Water, a state subject as per the provisions of the constitution of India, falls under the responsibilities of state governments for the supply and resource management (MPD 2021). For effective management of water supply and sewerage in the NCT of Delhi, in 1998, Delhi Government reconstituted the "New Delhi Water Supply and Sewage Disposal Undertaking" into the DJB, with the Chief Minister of the NCT as the Chairman of the Board.[2] The DJB, an independent agency under the state government, is responsible for the water resources management including production (and treatment), quality control, distribution, and monitoring. Wastewater collection, conveyance, treatment, and disposal facilities are also the responsibility of DJB (PWC et al. 2004). DJB is also fully responsible for all these services in the MCD area and supplies treated water in bulk to the NDMC and to the DCB, both of which are accountable for the distribution of this water within the respective territories (Figure 7.1).[3]

DJB's water resources are from, both, surface water and groundwater. River Yamuna, Western Jamuna Canal (a carrier of Yamuna waters as also Bhakra waters), and the Upper Ganga Canal, are major surface water sources for Delhi (Figure 7.2, Table 7.1).[4] Till 2001, most of the surface water was treated at the five plants: Chandrawal, Wazirabad,

Figure 7.1:
Broad Jurisdictional Territories of Water Supply Areas in Delhi

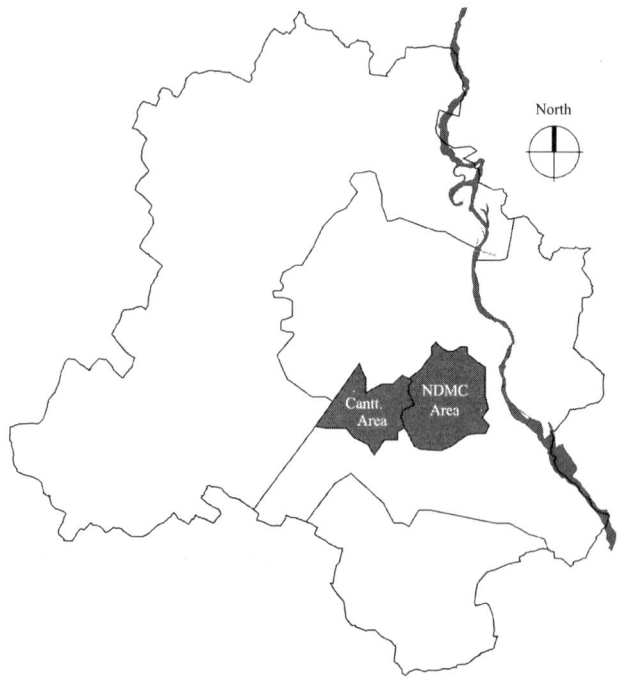

Source: Adapted from DJB.

Haiderpur, Bhagirathi, and Nangloi and since 2006, Sonia Vihar plant has also started functioning. To support the supply, groundwater has also been extracted through 16 wells (the "Ranney wells") at the Yamuna River channel bed along with about 2,760 tube-wells spread throughout Delhi (PWC et al. 2004). In 2001, total water treatment capacity was 650 million gallons per day (MGD), which, by 2021, is expected to rise up to 919 MGD (Table 7.1). However, that appears to be far short of the water requirements calculated by the DJB and the DDA (Table 7.2).

There are variations of about 50 meters between the highest and the lowest elevations in Delhi and the water sources are mainly at the lower levels of the Yamuna River and the canals supplying raw water (PWC et al. 2004). The situation demands the pumping of water from the treatment plants to the supply and booster stations have been provided in the system wherever required, as well as for the onward delivery in the distribution system. By and large, water supply system is through piped

Figure 7.2:
Water Treatment Plants in Delhi: Sources and Networks of Fresh Water

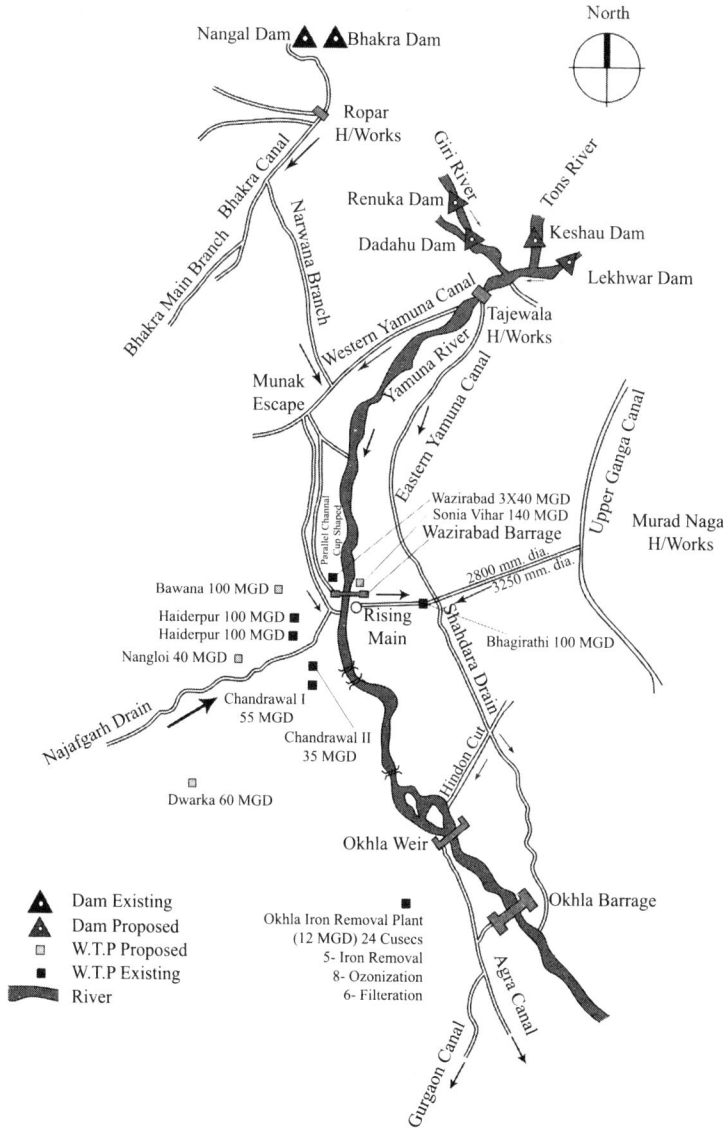

Source: Adapted from NCRPB 1999.

Note: W.T.P: Water Treatment Plant

Table 7.1:
Water Augmentation Plan of Delhi

Sources of Raw Water	Water Treatment Plants	Capacity 2001 (in MGD)	Capacity 2021[a] (in MGD)
River Yamuna	Chandrawal I and II	90	100
River Yamuna	Wazirabad	120	130
Bhakra Storage (I) and Yamuna (II)	Haiderpur I and II	200	216
Upper Ganga Canal	Bhagirathi	100	110
Upper Ganga Canal	Sonia Vihar	–	140
Bhakra Storage	Nangloi	40	40
Saving from seepage losses with the constructions of new lined channel	Dwarka	–	40
Do	Bawana	–	20
Do	Okhla	–	20
Sub-surface water	Ranney Wells at Okhla	100	12
Do	Palla and other ground water sources	–	91
Total		650	919

Source: Compiled by the Author from the data presented in DDA (2007a: 106, 141–145).
Note: [a]Capacity projected by the DJB.

Table 7.2:
Variable Estimation of Water Demand as per DDA and DJB Norms

Year	Water Demand (Treated) MGD[a]		Population (in lakhs)
	As per DJB at 60 GPCD[b]	As per DDA at 80 GPCD[b]	
2001	828	1,104	137.8 (as per census)
2011	1,140	1,520	190.0 (as per projection)
2021	1,380	1,840	230.0 (as per projection)

Source: Compiled by author.
Notes: Data source is DDA (2007a: 140–141).
[a]The raw water requirements will be 110 percent of the demand figure.
[b]1 GPCD (Gallons per Capita per Day) = 3.785 lpcd.

water network in urban areas, hand pumps and private motorized wells/ tube-wells, and in areas without planned water supply, through tankers.

Prevailing Discourses on Water Supply in Delhi

Discussions on the apparent shortcoming in water supply in Delhi give rise to a rich body of work and some of the established notions on the issue (Bhaduri and Kejriwal 2005, Dutta et al. 2005, Hoyt et al. 2005, Maria 2008, Parivartan ca. 2005, Singh and Shukla 2005, Zérah 2000a, 2000b).

Zérah (2000a, 2000b) proposes that the condition of unreliability, owing to the lack of round-the-clock supply and the insufficient water pressure, is an important pointer not only to understand the piped water supply in Delhi but also to assess consequences of unreliability on household behavior. Her study across Delhi highlights respective strategies adopted by individual household with in-house connections for attaining a degree of autonomy from the inadequate supply. Hoyt et al. (2005) mention the methodologies of data collection through public participation and the use of local knowledge system to bring about inclusive planning processes in delivering water services and infrastructure in informal settlements in Delhi. Likewise, attempts are made to reach empirical models, based on stakeholders' responses, for improving the quality and the reliability of water supply in unplanned settlements (Dutta et al. 2005, Savage and Dasgupta 2006). In a report sponsored by the WaterAid India, Singh and Shukla (2005) discuss the very existence of the informal settlements and their limited access to services, such as water and sanitation, within the backdrop of policies, urban development programs, and the institutional framework.

The mismanagement of the public utility, an economist and urban researcher Augustin Maria (2008) points out, is responsible for the insufficient and unreliable piped water supply by the DJB, and the inadequate supply is often supplemented by the groundwater privately extracted in Delhi. Maria (2008) also proposes the "World Bank model of the policy reform" involving the private operation and maintenance system as a possible alternative to the state-controlled practice of the urban water management system in Delhi that is presently dictated by norms, the demand–supply gap, and the augmentation of the source of water. The DJB has shown sufficient inclination to adopt the World Bank model of privatization as a strategy to improve water supply situations (DJB 2004, PWC et al. 2004). Such initiatives have met with severe criticisms of neoliberal restructuring and the resultant reshaping of the urban infrastructure networks (Bhaduri and Kejriwal 2005, Parivartan ca. 2005). These analytical positions are, in fact, in tune with the assessment of similar endeavors elsewhere (Bakker 2003a, Graham and Marvin 2001).

Most of the discourses tend to focus on slums and unauthorized colonies, the informal settlements, and, to a great extent, make gaps in the distribution and delivery of water supply by the DJB apparent.

The Gap between the Planned Distribution and the Actual Delivery of Water Supply

There are a number of discussions on the gaps prevalent between the planning and the actual delivery. Different sources cite different facts on the coverage of the DJB water supply. The disparity between the planned provision worked out of the intended demand–supply projection, and the real delivery of water supply indicates shortfalls in the coverage. The incongruent estimates of the non-revenue water (NRW) from various interest groups increase the doubt of coverage too.

Variable Facts on the Coverage of Water Supply

There are different views on the coverage of areas by the DJB's supply. The absence of information on the exact areas served by the distribution system does not clarify the issue and gets compounded by the "lack of record drawings" (PWC et al. 2004).

In 2000, Delhi's Planning Department, the National Commission on Population, Economic Survey of Delhi (2001–2002), and the DJB have reported about 100 percent coverage of water supply through both in-house connections and public stand posts. The following excerpt from the report on "Delhi Water Supply and Sewerage Project (DWSSP)" in 2004 gives an account of the coverage:

> Of the 135 urban villages, all have been supplied with piped water. Of the 567 unauthorized colonies 560 were supplied by March 2002, including 6,029 stand posts. The number of JJ colonies is not known, but 820 have

already been provided with piped water. There are 11,533 stand posts within the entire system (including 830 covering all markets). (PWC et al. 2004)

On the contrary, the National Sample Survey Organization (NSSO) observed in its 58th round of survey that despite the increase in the service outreach in Delhi by 14 percent from the previous (54th) round of survey, water supply could reach about 84 percent of the population only and, besides that, poor quality and erratic supply from taps and hand pumps installed in slums negated the significance of the infrastructure (Grover 2002, Hoyt et al. 2005). Interestingly, the Planning Department of the Government of NCT of Delhi suggested in the report, titled "An Appraisal of Annual Plan 2004–05," that the piped supply had been augmented in urban villages and unauthorized colonies, especially in *Jhuggi Jhopri* (JJ) clusters.[5]

It is hard to believe that, under such a condition, the coverage would be 100 percent.

Disparity between Planning (Intentions) and Implementation (Reality) in Water Supply

Indeed, differences exist between the projected and necessary requirements of water as well as the real availability of the provision of water. There are gaps in the projection and the necessary requirements: In 1981, necessary requirements were almost double than the projected one, whereas in 2001, it had been marginally less (about 3 percent) than that in the latter year. Availability of water resource, in 1981, was almost the same as the projected requirement but, in 2001, it had been far less (42 percent) than that in the latter year and in 2021, additional 500 MGD of water is to be made available (Figure 7.3). In 1981, planning projection had gone terribly wrong in considering the number of people to whom the resource was to be distributed and in 2001, in ensuring how much of resource is available for distribution.

Shortfalls in demand–supply projections, too, are evident. As per the MPD 2021 by the DDA, projected water supply for 23 million populations would be 1840 MGD, whereas DJB's Capacity Augmentation Plan 2021, falling short of that target, indicates 919 MGD of water from all possible sources (DDA 2007a; Tables 7.1 and 7.2). In 2004, domestic

Figure 7.3:
Availability and Projections for Water Supply in Delhi

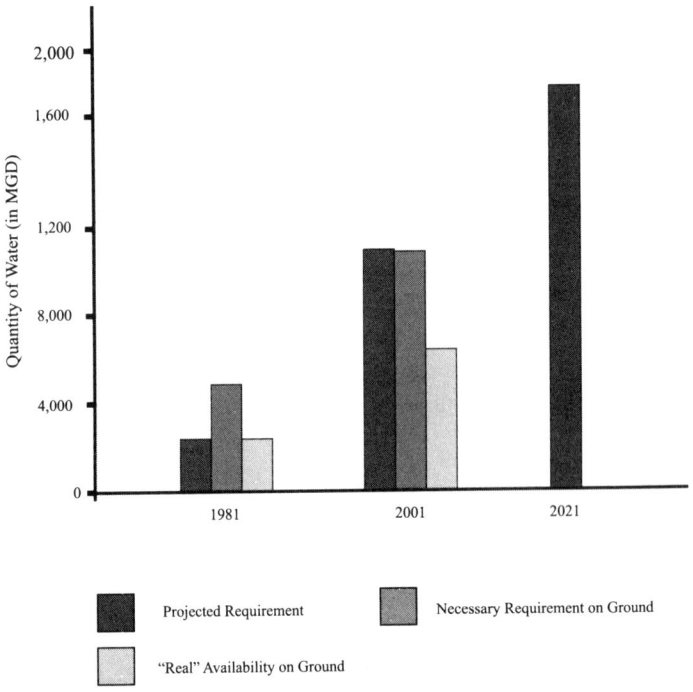

Source: Author.
Notes: 1. Data source is DDA (1990, 2007a).
2. The projected requirement is based on the population projections. The necessary requirement is the quantity required as per the actual population figure and the real availability on
ground is the quantity of water that can be supplied.

water demand had been 456 MGD (2,076 thousand cubic meters per day
[TCMD]), and in 2001–2002, the sale of water for domestic purpose was
247 MGD (1.124 TCMD), about 209 MGD less than what the demand
would have been (PWC et al. 2004). The total domestic water supply has
been much less than the projected one.[6]

Differences regarding the policy norms for the per capita demand exist
across government agencies. Domestic daily per capita water demand as
per the DJB records (121 lpcd) has been 20 percent less than the figure
(145 lpcd) of the "Willingness to Pay" survey conducted during a study,
commissioned by the DJB on water supply in Delhi (PWC et al. 2004).
However, the MoUD of the Government of India in its *Handbook*

of Service Level Benchmarking (ca. 2008) identifies 135 lpcd as the benchmark for the supply norms. MPD 2021, following the CPHEEO Manual (1999) on the norms for water supply in the metropolitan and mega cities, suggests the quantity of 172 lpcd for domestic consumption, calculated by including a loss of 15 percent on top of 150 lpcd (DDA 2007a). Interestingly, Annexure B of the MPD 2021 document also includes a later estimate by the DJB showing the total of 225 lpcd (with a break-up of 135 lpcd for potable water and 90 lpcd as non-potable water) (DDA 2007a: 148).[7]

Water supply seems adequate in a residential colony, whereas the adjacent colony does not enjoy the same, which somewhat gives a sense of spatial inequity of supply across the city. A newspaper reporting during the summer season mentions;

> Rohini and its adjoining area in West Delhi have problems of inadequate water supply and low pressure despite closeness to the Haidarpur water treatment plant. ... South Delhi situation is no better. Munirka residents complained about *inadequate supply and low pressure* and mentioned that they use bore wells. Saket residents complained about *the irregularity and bad quality of supply.* (Times News Network 2004; emphasis included)

This kind of news on water supply in Delhi is quite common, especially in summer months. MPD 2021, too, acknowledges this point. Similar news-items for other services are also found almost every day, in particular for the power situation in summer. The detail report, commissioned by the DJB, mentions that,

> [T]he supply of water in the service area is far from even. Though half the revenue zones have an average supply per connection of between 25 and 35 cubic metres per month (m3 per month), six zones have more (up to 73 m3 per month) while eight have less (as low as 2 m3 per month). To provide water to the areas not covered by the distribution system, DJB supplies water by tanker service. (PWC et al. 2004)

The vision of the MPD 2021 recognizes a considerable gap existing between the two situations of the basic services, the planned provisions and the delivery on ground (DDA 2007a). Regarding the availability of basic infrastructure facilities and their accessibility, both required for attaining a minimum quality of living, the MPD 2021 underlines "the need for capacity building, 'User Pays' approach and PPP as tools for institutional strengthening" and identifies the role of "community participation and decentralized management" to improve the efficiency and the performance of basic services (DDA 2007a: 104). DDA also mentions in the *MPD 2021—Vision Plan* about the shortcomings found in previous plans:

> The experience of the past two Master Plans also shows that while projections regarding various basic infrastructure services have been made with reference to the population growth projections and the related increased urbanization requirements, *there has been very little practical convergence between the Master Plan and the actual development of infrastructure services.* An important element would, therefore, have to be brought in to bring greater convergence between these two aspects, particularly in the areas, which would be taken up for fresh urbanization. (DDA 2007a: i–v; emphasis included)

How far the MPD 2021 will be able to mitigate the gap by bringing greater convergence between planning and implementation is to be seen in the due course.

Ambiguity Surrounding the Non-revenue Water

The demand–supply gap gets even wider due to the *loss of water* in the distribution process (Figure 7.4). *Loss of water* in distribution or in transit is often referred to as "unaccounted for water" (UFW) and "NRW" as well. According to the PWC report, the best practice figure for the loss of water would be below 20 percent (PWC et al. 2004). However, different figures ranging from 24 percent to 45 percent are available on the amount of the loss of water in Delhi. PWC report, funded by the DJB, indicates loss of water in the distribution being high, about 24 percent (PWC et al. 2004). National Institute of Urban Affairs (NIUA) report (2005) based on the data in 1999, mentions 26 percent UFW. Percentage of UFW, calculated from the difference between water produced and water pumped, is in the range of 35–40 percent thereby reflecting problems in the management of available resources (Singh and Shukla 2005: 36). Delhi Urban Environment and Infrastructure Improvement Project (DUIIP) report suggests 40 percent loss of water in transit (Ministry of Environment and Forests 2001: 39). Economic Survey of Delhi 2001–2002 finds out 44 percent UFW in Delhi (Government of NCT Delhi 2001–2002: 116). More recently, a newspaper article suggests a loss of 45 percent water in transit (Lalchandani 2009: 4).

Figure 7.4:

Loss of Water in Transmission, as We Witness Everyday

Source: Devneil Biswas (2011).

An analysis by a Delhi-based NGO, Parivartan (ca. 2005), of the report on Delhi Water Supply and Sewerage Project prepared by the PWC et al. (2004), reveals that even at the zonal levels, for example South II and South III zones of the DJB where the privatization pilot scheme has been proposed, estimates of the NRW vary considerably across various public and private agencies: Estimates by the DJB, the PWC, and the GKW are 24 per cent, 48 percent, and 59 percent, respectively. DJB targets to reduce NRW from 55 percent to 34 percent in three years, which, as per international precedents, appears highly improbable (Parivartan ca. 2005).[8] The ambiguity surrounding the variable estimates of the NRW from different interest groups somewhat suggests a sort of redundancy of the projected NRW targets by the DJB.

All these observations indicate that water supply is a scarce urban service in Delhi. Also, quantitative gaps exist between the intended policy and the real delivery, which, over time, have been growing wider. As a result, the state, in desperation, looks for newer and often further located water sources leading to what Karen Bakker (2005), a leading expert on water governance, terms as the "state hydraulic" model of water management. The apparent demand–supply gap and the resultant shortfall of coverage of supply are expected to lead to the distribution of insufficient quantity of water in general.

Within such differences in the demand–supply that make urban water supply a scarce service, how is the equitable distribution, envisaged in the welfare planning model, addressed in the city? Since the distribution of water to residential colonies is one of the responsibilities of the DJB, which is the public body under the state government, one may expect that at least the state, within a democratic set-up, would ensure an equitable distribution and delivery of water across colonies. But on ground, water supply varies spatially from colony to colony.

The subsequent perception is that the distribution of water in Delhi is inequitable across spaces and people—but how, and is there any *pattern* to it? My observations are essentially threefold:

- First, there are clear evidences of the inequitable distribution of water across broad geographical spaces in the city.
- Second, inequity out of differential policy norms is observed across various settlement types (and the land tenure) in the city.
- Finally, I shall argue based on the primary empirical results that even among the planned colonies, water supply is higher for higher socioeconomic spaces in Delhi.

Horizontal Inequity across Geographical Spaces in the City

Inequity at the broad spatial level across Delhi had been identified when the NCRPB presented a map in 1999. Some of the key findings of that map are: The level of supply is the highest in the Cantonment area at 509 lpcd—almost 18 times the level of supply in the Mehrauli area; Narela and Mehrauli, the peripheral areas of the city, have very low levels of water supply at 31 and 29 lpcd, respectively; and the level of supply in South Delhi is low (148 lpcd) considering the high demand from a largely medium-high-income residential area (NCRPB 1999; Figure 7.5).

The overall low level of supply in Narela and Mehrauli seems predictable as these areas include a large number of rural villages having the supply norm less than that for planned colonies. However, a very low level of 29 and 31 lpcd cannot be justified for any part of Delhi. In

Figure 7.5:

Water Supply (in lpcd) in MCD Zones, NDMC, and DCB Areas in Delhi

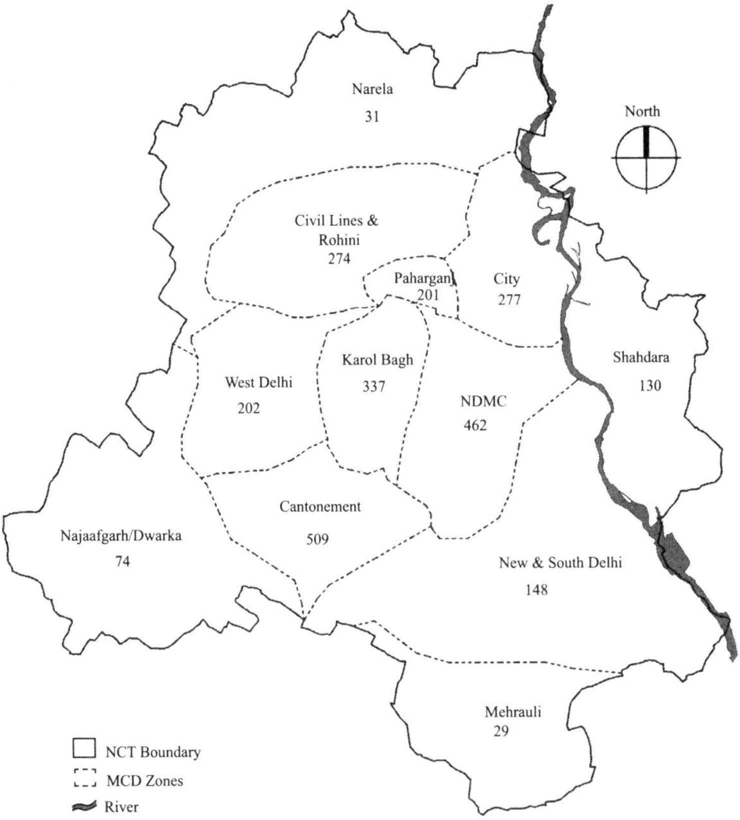

Narela
31

North

Civil Lines &
Rohini
274

Paharganj
201

City
277

Karol Bagh
337

Shahdara
130

West Delhi
202

NDMC
462

Najaafgarh/Dwarka
74

Cantonement
509

New & South Delhi
148

Mehrauli
29

☐ NCT Boundary
⌐ ⌐ MCD Zones
〰 River

Source: Adapted from NCRPB (1999).
Note: Optimum Supply: 150–225 lpcd.

2001, the DUIIP report, jointly prepared by the Government of India and the Government of NCT, too, has recognized that some areas received quantities of water far in excess of the policy norms, whereas 10 percent of the population had no piped water supply and 30 percent had grossly inadequate access to safe drinking water (Ministry of Environment and Forests: GoI and GNCTD 2001: 39).

Often, views within most of the government institutions, albeit Malthusian in notion, are that the water supply infrastructure has failed to match the demand because of the rapid and unexpected urban development, especially the unplanned or uncontrolled growth of the city (Ministry of Environment and Forests 2001). Consequently, ground water resources have been exploited in places, which, in turn, became depleted or saline and many areas of the city, where the "urban poor" lives, now have to rely on water tankers (ibid.). It is also reported that along with the slum areas, many South Delhi colonies, too, live off the tankers (Navdanya 2005). It indicates spatial inequities at the city scale apart from differences in water supply due to the settlement types or the economic status of the colonies.

Five spatial zones in Delhi may be identified on the basis of water supply conditions: The first, where treated piped water is available for 24 hours; the second, where rationed water is available for a total of 6 hours a day; the third, where tube-well water mixed with inadequate piped water supply is provided for few hours; the fourth, where water, untested for its quality, is made available only through tube-wells and hand-pumps; and the fifth, where no organized water supply is made available (Ruet et al. 2002, Susheela et al. 1996: 356).

Also, the CDP of Delhi mentions that the availability of water is for 2–3 hours in general and in some areas, mostly in North Delhi, the supply is directly on line for 24 hours, thus pointing to an inequitable distribution of water in different parts of Delhi (IL&FS Ecosmart Ltd 2006). The supply of water is reported to be *far from even* across OZ of the DJB as well. In half of these zones, average supply per connection has been between 25,000 and 35,000 liters per month and six zones have more supply, up to 73,000 liters per month, while eight have as low as 2,000 liters per month (PWC et al. 2004). DJB supplies water by the tanker service to the areas not covered by the piped distribution system (ibid.). Price WaterHouse report in 2004 also identifies a couple of obvious reasons behind the uneven distribution of water: On the one hand, due to direct connection to the transmission mains in a 24-hour operation, number of distribution zones receive almost round-the-clock supply and on the other hand, due to "direct tapping" to the transmission system, the flow of water varies significantly during the day, somewhat indicating the hour-based demand in the areas fed by the direct supply (ibid.). However, these issues are more relevant in a round-the-clock supply system. In reality, with inadequate supply (quite often 2–4 hours of supply in a day) across most of the areas in the city, hourly demand variations do not mean much. Majority of the areas apparently do not even have the luxury of having longer hours of water supply, except where water is sold to other authorities, such as the Cantonment and the NDMC areas—the military and the administrative/ political establishment, respectively.

Researches also suggest that with the gradual deterioration of the water situation, over 35 potential "trouble areas" across "up-market," "middle class," and "LIG" colonies have been identified by the Delhi Police and to diffuse any potential situation, the local police would have to coordinate with the RWA (Navdanya 2005: 61).[9] Interestingly, quite a few localities in South Delhi, the chosen area for the empirical study, suffer from the "scarcity of water supply" (ibid.). One of the reasons for the inadequate supply in the southern part of the city, the Economic Survey of Delhi (2001–2002) points out, can be the location of the water "production centers," namely Haiderpur, Wazirabad, Nangloi, and Chandarwal, in the northern part of Delhi, because of which the trunk mains have had to carry water long distances resulting in the loss of pressure and flow and at the same time, the distance increased the susceptibility to the "illegal tapping" (ibid.).

All these observations, indeed, reveal the horizontal inequity across geographical spaces in the city. However, one may also notice further inequities in supply within a smaller area, like South Delhi, contributed by differential policy provisions.

Inequity Out of Differential Policy Norms

The mainstream discourses on the inequitable distribution of water in Delhi have prevailed upon the difference in institutionalized *norms* for water supply in colonies according to the land tenure or the settlement types (Hoyt et al. 2005, Maria 2008). Delhi has a number of seven to eight categories of housing stocks, often mentioned as types of settlement in official planning documents (Banerjee 2002, Dhar Chakrabarti 2001, Maria 2008, The Urban Institute 2007).[10] The provision for per capita water supply is stipulated by the government agencies with an assumption that different segments of population have different (and specific) demands of the quantity of water (Table 7.3).

The average per capita norms, too, vary across various reports and agencies creating a lot of ambiguities. In the MPD 2021, DDA recommends a supply norm of 225 lpcd, which is much lower than the MPD 2001 recommendations of 363 lpcd. DJB, on the contrary,

Table 7.3:
Differential Policy Norms for Different Settlement Types

Type of Settlement	Population in 2000 (in million)[a]	% of Total Population[a]	Tenure[d]	Poverty[c]	Access to Individual Connection[d]	lpcd[d]	Supply Norms[e]
JJ/Slum clusters (Squatter Settlements)	2.072	14.8	Illegal	High	No right to individual connection	70	CPHEEO Norms
Slum designated areas	2.664	19.1	Legal	Mixed	Restricted by technical features	70	CPHEEO Norms
Unauthorized colonies	0.740	5.3	Semi-legal	Mixed	No right to individual connection	70	CPHEEO Norms
Resettlement colonies	1.776	12.7	Legal	High	Official right not respected	150[a]	Based on CPHEEO Manual
Rural villages	0.740	5.3	Legal	Low	Not under the responsibility of the DJB	150[a]	Based on CPHEEO Manual
Urban villages	0.888	6.4	Legal	Mixed	Good situation	168[a]	Based on MoUD norms
Regularized unauthorized colonies	1.776	12.7	Legal	Mixed	Good situation	168[a]	Based on MoUD norms
Planned colonies	3.308	23.7	Legal	Low	Good situation	225[b]	Based on MPD 2021
NCT of Delhi	13.964	100.0					

Source: Compiled by author.

Notes: Minimum per capita demand is maintained as 70 lpcd. As per Water Aid report, standard for resettlement colonies and urban villages is 155 lpcd and for JJ Cluster, it is abysmally low at 50 lpcd (Singh and Shukla 2005). Break-up of 225 lpcd for Planned colony (135 lpcd potable water + 90 lpcd non-potable water (DDA, 2007a)

[a]Government of NCT Delhi. (2004)

[b]DDA (2007a).

[c]Jain (1990).

[d]Maria (2008).

[e]Price Waterhouse Coopers, GHV, TCE (2004).

recommends 131 lpcd. Apart from that, the NIUA report in 1999 observed that the city level water supply was 218 lpcd and the slum dwellers generally received about 25 lpcd of water. In 2002, the *Delhi 21* report measured the supply to one-third of the city as between 4 and 10 lpcd through stand posts (Hoyt et al. 2005). In 2005, a report on the informal city of Delhi by an NGO, WaterAid India, highlighted that despite the norms for informal housing being 40 lpcd and 1 community stand post for 150 persons, the actual provision in informal settlements was about 30 lpcd (Hazards Centre 1999, NCRPB 1999, Singh and Shukla 2005). However, the Environmental Improvement in Urban Slums (EIUS) scheme of the Government of India recommends one tap per 50 people as the norms for the community level stand posts (Hoyt et al. 2005).

At the city level, calculations for the per capita requirement and the overall water demand, both, tend to get faulty owing to the wavering number of the *informal* settlements considered by various sources (Table 7.4). It appears that the DJB's responsibility to provide individual water supply to settlements with 43 percent of the total population is somewhat clear, but around 20 percent of the population, living in JJ clusters and unauthorized colonies toward the other side of the spectrum, have no right to individual water connections (Table 7.3). Instead, they receive less reliable services at community points and slum dwellers are dependent on hand pumps installed by the DJB. Water from 40 percent of these hand pumps is unfit for drinking (Hoyt et al. 2005: 5). At the same time, the politics of difference in the number of *informal* settlements would obviously affect the population estimates and the corresponding supply (Table 7.4).

In resettlement colonies, where provision of basic services up to the plot would be expected, the access to the piped water supply and

Table 7.4:
Number of Informal Settlements in Delhi

Type of Settlements	No. of Settlements				
	NIUA (1991)	*DDA (1994)*	*NSSO (2002)*	*Delhi Planning Board (2003)*	*DJB (2004)*
JJ Clusters (Slum Settlements)	1,190	1,080	1,847	1,080	820
Resettlement Colonies	–	–	–	44	44
Unauthorized Colonies	–	–	–	1,017	1,017
Urban and Rural Villages	–	–	–	354	345

Source: Hoyt et al. (2005).

sewerage is reported to be low in most of these rather recently established colonies (Ali 1995, Hazards Center 2004). The Planning Department of the Government of NCT of Delhi mentioned in the outline of physical targets and achievements of *An Appraisal of Annual Plan 2004–2005* that about 98 percent unauthorized regularized colonies have been provided with piped water supply system and the piped supply to the unauthorized colonies and to the JJ clusters have been augmented along with the increase in other provision for water, for example, tanker supply, stand post, etc. (Government of NCT of Delhi 2004–2005: 110–112). DJB records from 2002 to 2005 also showed consistency in the allocation of funds and indicated the spending of about ₹5,000 lakhs on water supply in the resettlement colonies, urban villages, and unauthorized colonies and JJ clusters; in 2006–2007, about 15.5 percent of the total budgetary provisions, that is, ₹6,860 lakhs, were also kept for these settlement types (DJB 2006–2007; Table 7.5).[11]

Indeed, the policy of such water supply targets at the per capita level, based on the land tenure status or the settlement types comprehended by the Master Plan, expose the embedded notion of the differential distribution within the rationality of allocation itself. This is an example of what Jessop (1990) recognizes as the "strategic selectivity" of the state. Also seen are the multiple techniques of governmentality in creating vagueness surrounding the information on the number of "informal" settlements and the per capita calculation of water requirement. The fund allocation in the budget is a practice usually followed by the government to indicate its attention to the issue of water provisions in informal settlements, which, however, is found unconvincing by many works by the NGOs and scholars.

The politics of information, here, is the manipulation of unclear ground realities out of the "hybrid mix" of rights, the "special treatment rights, text-based rights and contributor rights," to live in the city leading to a combination of legal and *para-legal* status (Chatterjee 2004, Holston 2008: 256, quoted in Bhan 2009: 133). The policy of differential allocation compounded by the governmentality of the multiple patterns of delivery of water (for example, supplying less water or installing less number of community points, etc., for the redistribution of the scarce source of water or to reduce the costs) to these para-legal arrangements can also be understood as outcomes of what Agamben (2003[2005]) terms as the "conditions of necessities". Ironically, this politics of distribution and delivery reveals another condition of necessity for the people: To supplement the lack of provision of water supply by the state,

Table 7.5:
Summary of the Annual Plan for Water Supply and Sanitation (2006–2007)

No.	Water Supply Schemes	Financial Outlay (lakhs)	Percent
1	Providing water supply in unauthorized colonies	4,100	9.3
2	Replacement of old distribution system and strengthening of the trunks transmission network	1,200	2.7
3	Improvement of existing water works	1,400	3.2
4	Ranney wells and tube-wells	4,400	9.9
5	Staff quarters and office accommodation	590	1.3
6	Laying of water mains in regularized/ unauthorized colonies	1,100	2.5
7	Raw water arrangement for additional. needs	10,005	22.6
8	Rural water supply	620	1.4
9	Distribution mains and reservoirs for second 100 MGD WtP Haiderpur	2,600	5.9
10	140 MGD WtP at Sonia Vihar	11,000	24.9
11	Construction of 40 MGD WTP at Nangloi	120	0.3
12	Construction of 20 MGD WTP at Bawana	185	0.4
13	Construction of 60 MGD Plant at Bakarwala (Dwarka)	60	0.1
14	Construction of 40 MGD water treatment plant at Okhla	20	0.0
15	Hydraulic mapping and study of water supply system	100	0.2
16	Augmentation of water supply for urban villages	480	1.1
17	Augmentation of water supply in resettlement colonies	400	0.9
18	Augmentation of water supply for JJ clusters	780	1.8
19	IT infrastructure	100	0.2
20	Trans-Yamuna Area Development board	1,500	3.4
21	Special scheme for Grant-in-aid for development of sewerage and water supply in ACs	3,500	7.9
Total		44,260	100

Source: Author.
Note: Data source is DJB (2006–2007); available at http://delhiplanning.nic.in/reports/ PDF/Water.pdf (accessed June 30, 2006).

people living in the notified slum areas in South Delhi tend to pay a "poverty premium" of four to seven times more than the DJB's price of water (Dimri and Sharma 2006).[12]

Governmentality of such varied dimensions of the inequitable distribution of water, may, perhaps, be regarded as the exposition of the rationalities or mentalities of the government in managing its affairs, which prime facie favors the *legal* status of the settlements in Delhi. I will now argue with the help of the empirical findings that, besides being connected with the settlement types, inequity in the delivery of water supply varies even with the socioeconomic spaces of the so-called *legal* colonies in Delhi.

Vertical Inequity in Delivery across Socioeconomic Spaces of Legal Colonies

The central assumption is that the legal property rights do not necessarily ensure equal water supply in socioeconomic spaces with same water supply norms and, in turn, higher strata of socioeconomic spaces get higher supply of water.

Since, the planning policy recommends same quantity of supply for colonies with legal property status, it is worth introspecting whether real delivery of such recommendations happen on ground. I address the assumption from two sets of empirical data: one, the supplier-end information at the colony level collected from the DJB (the DJB data), and the other, the user-end information collated from a sample survey during 2006–2007 (the survey data).[13] These colonies represent a range of socioeconomic spaces indicated by all four types of strata, covering property tax categories from A to G.

The DJB data represents the supplier-end information at the colony level for 48 colonies, having a total population of about 6.10 lakhs, covered by three major UGR in South Delhi, namely GK 1, Giri Nagar, and Jal Sadan. This dataset clearly points out that higher the tax strata, that is, the higher the socioeconomic strata, higher the per capita quantity of the filtered water supply and that of the total water supply. Also, more number of maintenance staff is engaged for per 10 thousand populations of the higher strata and as per the annual operational expenditure in 2006 by the DJB, maximum money has been spent in the highest strata (Table 7.6).[14]

Table 7.6:
Relationship of Tax Strata with Water Supply Based on DJB Data

Strata Based on Property Tax Category	Estimated Marginal Mean			
	Quantity of Filtered Water Supply (lpcd)	*Quantity of Total Water Supply (lpcd)*	*No. of Maintenance Staff per 10,000 People*	*Annual Operational Expenditure by the DJB for per 100 People Served in 2006 (₹)*
I	152.70	161.23	19.76	3,127.90
II	119.43	133.25	13.89	2,228.70
III	101.36	117.51	11.11	2,471.10
IV	95.71	105.18	10.88	2,447.70
Grand Mean	117.30	129.30	13.91	2,568.80

Source: Author.

The key variable of the supplier-end information is, indeed, the quantity of filtered water supply (in lpcd) indicating the actual delivery, whereas the maintenance staff and the operational expenditure signify the resource allocation in terms of overheads. There are significant correlations ($p < 0.05$) among these four factors of water supply; especially the variable "Quantity of Filtered Water Supply (lpcd)" is significantly correlated with other four variables ($p < 0.05$). Efficiency of the supply, for example, the complaint redressal by the authority, could not be measured either through the DJB data or the survey data. Similarly, variations in pumping hours in the DJB information, too, seem quite insignificant, which, otherwise, could have indicated the system reliability.

Average quantity of the filtered water supply (in lpcd) is clearly higher for the higher strata and lower for the lower strata (Figure 7.6). Average filtered water supplied (117.30 lpcd) seems comparable with the average supply standard as per the DJB records (121 lpcd), but is over 12 percent short of the recommended standard of the CPHEEO/MoUD (135 lpcd). Only the topmost strata, that is, Strata I, enjoys the filtered water supply around 13 percent more than the recommended standard of 135 lpcd. Strata I and II have supply above the overall average supply of the filtered water (117.30 lpcd), whereas Strata III and IV have supply far below the same, thereby creating two clear slabs/divisions across socioeconomic spaces (Table 7.6).

ANOVA results show significant effects of the strata on the filtered water supply (in lpcd) [$F(3, 44) = 5.666$, $p < 0.01$], on the total water

Figure 7.6:
Strata-wise Average Quantity of Filtered Water Supply (in lpcd)

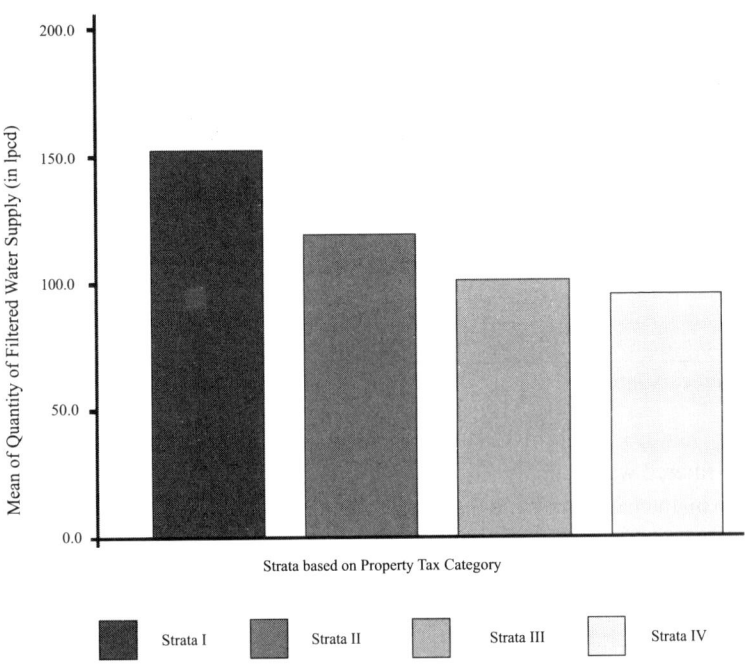

Source: Author.

supply (in lpcd) [$F(3, 44) = 5.312, p < 0.01$], and on the annual operating expenditure in 2006 [$F(3, 44) = 6.224, p < 0.01$], whereas the strata has a partially significant effect on the number of maintenance staff, engaged per 10,000 people in different strata [$F(3, 44) = 2.403, p = 0.080$].

Factor analysis is also adopted to observe composite effects of all four variables, namely quantity of filtered water supply (in lpcd), total quantity of water supply (in lpcd), maintenance staff per 10,000 people, and annual operating expenditure in 2006 for the distribution of filtered water per 100 people (₹).[15] Hence, the factor of water supply (FWS as per the DJB data) is chosen as the parameter for water supply for further analysis. After extraction, FWS (Component 1) has significant correlation with all four variables ($p < 0.001$).[16] Estimated mean value of FWS is clearly more for the higher strata and less for the lower ones and the Strata I is the only one having mean value higher than the grand mean [= (−)0.132] (Figure 7.7, Table 7.7).

Figure 7.7:
Strata-wise Average FWS

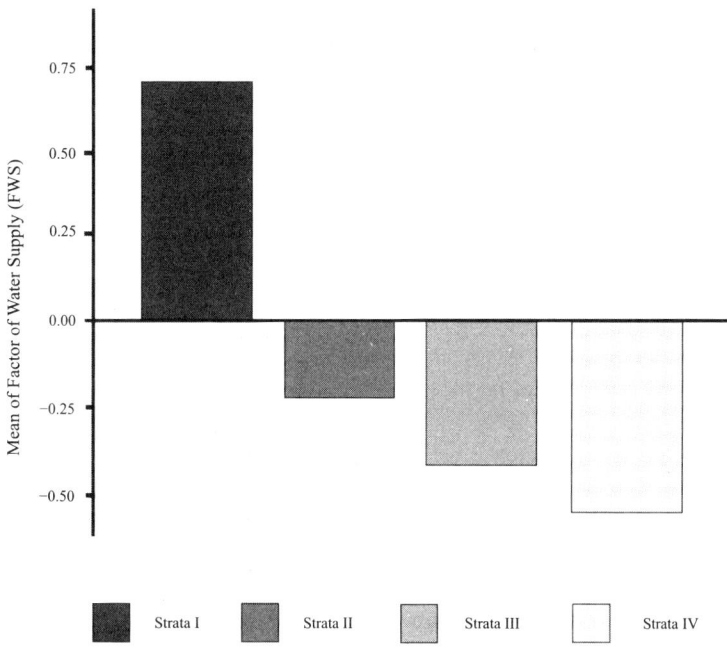

Source: Author.

Table 7.7:
FWS as per DJB Data and Tax Strata: Relationship-I

Strata Based on Property Tax Category	Mean	Std. Error	95% Confidence Interval		Bootstrap for Mean[a]			
			Lower Bound	Upper Bound	Bias	Std. Error	95% Confidence Interval	
							Lower	Upper
I	.709	.211	.284	1.134	−.001	.185	.351	1.075
II	−.231	.262	−.760	.297	.005	.335	−.850	.449
III	−.433	.232	−.901	.035	−.008	.229	−.864	.048
IV	−.574	.355	−1.290	.141	−.006	.248	−.946	−.011
Grand Mean	−.132	.135	−.405	.140	−.002	.129	−.381	.116

Source: Author.
Note: [a]Unless otherwise noted, bootstrap results are based on 1,000 bootstrap samples.

To determine whether differences in water supply are affected by the strata, data are subjected to a univariate ANOVA with FWS as the dependent variable and the strata as the independent variable. The bootstrap sampling method is used, showing a significant effect of the strata on the FWS, $[F(3,44) = 6.055, p < 0.01]$.[17]

The survey data represent the user-end information, obtained from the sample survey conducted during 2006–2007 over 1,477 households in 63 colonies across 17 survey clusters. Analysis of the data gives another dimension to understand the delivery of water supply. Surveyed colonies include households from all property tax zones from A to G covering all four strata.

Interviewee profile comprises of 39 percent female and 61 percent male population. Majority of them (93.2 percent) are within 18–60 years of age. Most of the interviewees (70.5 percent) are owners of their property and, hence, permanent residents, whereas 29.5 percent of the interviewees stay in rented houses. The number of families with over four members is high (Table 7.8). The interviewee profile is representative of gender, age, and property rights. In fact, the presence of a larger number of owners in the profile should ensure better information on the colonies.

DJB piped supply is the primary source of water for the households surveyed (Table 7.9). About 80.4 percent households have the DJB supply among which 61.9 percent have water only from the DJB supply and the rest 18.5 percent have a second source of water either from the tankers or from the bore well. A large number of households, about 36 percent, in the Strata I have bore well as the second source of supply. Here is an indication of affordability easing out the deficit in the DJB supply, which, to an extent, supports the idea of autonomous arrangements for unreliable supply, suggested by Zérah (2000a, 2000b). Thirty-nine households, which do not fall within the property tax categories

Table 7.8:
Variations in Family-size across Surveyed Households

Family Size	Percentage (Valid N=1,466)
Less than two members	1.0%
2–4 members	61.7%
More than four members	37.3%
Mean family size	4.32
SD	1.71

Source: Author.

Table 7.9:
Sources of Water Supply for Surveyed Households

		Frequency	Percent	*Valid Percent*	*Cumulative Percent*
Valid	Tanker	38	2.6	2.6	2.6
	Bore well	251	17.0	17.0	19.6
	DJB-piped only	914	61.9	61.9	81.5
	DJB-piped + Tanker	37	2.5	2.5	84.0
	DJB-piped + Bore well	236	16.0	16.0	100.0
	Total	1,476	99.9	100.0	
Missing	System	1	.1		
Total		1,477	100.0		

Source: Author.

(and grouped under Strata Z), have no DJB supply. Average expenditure per household across all strata for DJB water is about ₹370.00. Average expenditure for the DJB water (₹489.38 per household) is the highest in Strata I which is indicative of more consumption of water.

The survey looks at two broad parameters of the quantity of water, namely the frequency (days/week) and the duration of supply (hours/day), and both are observed for summer months (May–July) and for the rest of the year. However, the frequency of supply, somehow, does not indicate much variation in terms of the number of days of supply per week, or the average frequency of supply during summer months and that during the rest of the year, or across the strata.[18] On the other hand, average duration of supply in the dataset in summer months is about four hours, almost an hour less than the rest of the year's average, which seems considerable. Hence, variables on the duration of supply are taken for factor analysis.[19] ANOVA results show that the strata has a significant effect on the factor of duration of supply (FDS) $[F(3,1067) = 30.375, p < 0.001]$. Estimated mean of the FDS over the strata show that the mean value of Strata I and Strata II are higher than the grand mean and Strata III and Strata IV have mean value less than the grand mean $[=(-)0.086]$, thereby creating two distinct slabs of supply (Table 7.10).

Interviewees were asked to rate, in a scale of 0 to 5, their level of satisfaction (LoS) on certain quantitative aspects of supply, namely quantity, regularity, and duration of supply. The factor, here, indicates

Table 7.10:

Estimated Marginal Means of FDS

Strata Based on Property Tax Category	Mean	Std. Error	95% Confidence Interval	
			Lower Bound	Upper Bound
I	.080	.046	−.010	.169
II	.289	.054	.183	.395
III	−.481	.066	−.609	−.352
IV	−.233	.097	−.422	−.043
Grand Mean	−.086	.034	−.153	−.019

Source: Author.

the combined LoS scores for three variables, namely quantity of water, regularity of supply, and duration of supply. Since all these variables are related to quantitative aspects of water supply, the factor is considered for measuring the user's LoS regarding the quantity of water supply.

The analysis shows that the middle and lower-middle socioeconomic groups are relatively less satisfied with the DJB supply, whereas both ends of the group seem more content with the quantity of DJB supply. Interestingly, the LoS for the quantity of DJB piped water supply is the highest at the topmost tax category, A. Reasons for two extreme ends of the spectrum, Strata I and IV, having similar higher LoS with the quantity of water supply may be related to the divergent sources of supply. In Strata I, there is higher level satisfaction for the DJB supply, whereas in Strata IV, that is for the non-DJB sources of supply, for example, bore well or tanker (Figures 7.8 and 7.9).

There is a significant effect of strata on users' LoS with the quantity of supply [$F(3,1118) = 24.145, p < 0.0001$]. Estimated mean of the factor of the LoS over the strata, too, shows similar patterns: higher values of LoS in Strata I and IV and lower ones in Strata II and III. Tax categories A, B, and G have higher mean values, whereas tax categories C, D, E, and F have lower mean values than the grand mean (=0.058). Similarly, Strata II and III have lower mean values than the grand mean (=0.035). It shows that people from middle and lower-middle level tax categories are satisfied least with water supply (Figure 7.10). Post-hoc tests reinforce observations mentioned before: Strata I and IV have significantly higher effect than II and III on the LoS of the users ($p < 0.0001$). However, inter-strata differences between I and IV ($p = 0.550$) and between II and III ($p = 0.445$) are not significant at 0.05 level. All these show that the socioeconomic spaces have significant relationships with the LoS with the water supply.

Figure 7.8:
User's LoS for the Quantity of Water Supply with Respect to the Source of Supply-I

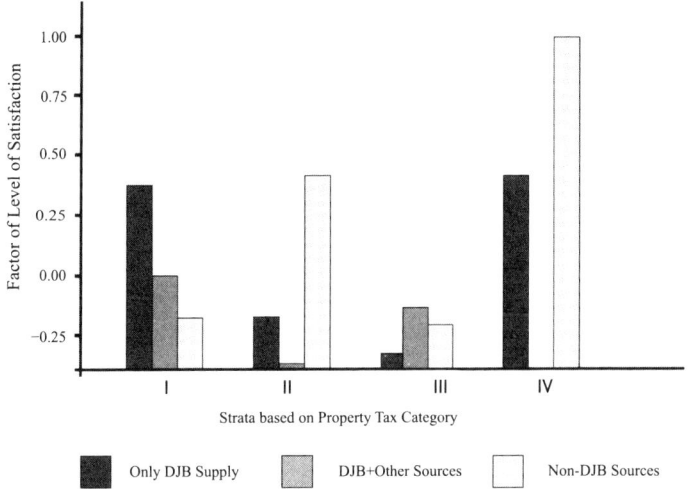

Source: Author.
Notes: 1. DJB sources mean the piped water supply.
2. DJB + other sources represent piped supply augmented with bore well.
3. Non-DJB sources of supply include only bore well and tanker in the absence of piped supply.

Factor of LoS is the combined score of three variables, namely quantity of water, regularity of supply, and duration of supply. Visual analysis of the data on the source of supply reveals three distinctive patterns in the mean value of the LoS regarding quantity of DJB water supply:

- LoS decreases from Strata I to Strata III and then increases to Strata IV. It also decrease from higher tax category of A to higher-middle category of D;
- LoS remains similar at higher-middle category of D to lower-middle category of F. Similarly, difference in the mean value of LoS in Strata II and III is marginal; and
- LoS increases at the lower tax category of G and at Strata IV.

Figure 7.9:
User's LoS for the Quantity of Water Supply with Respect to the Source of Supply-II

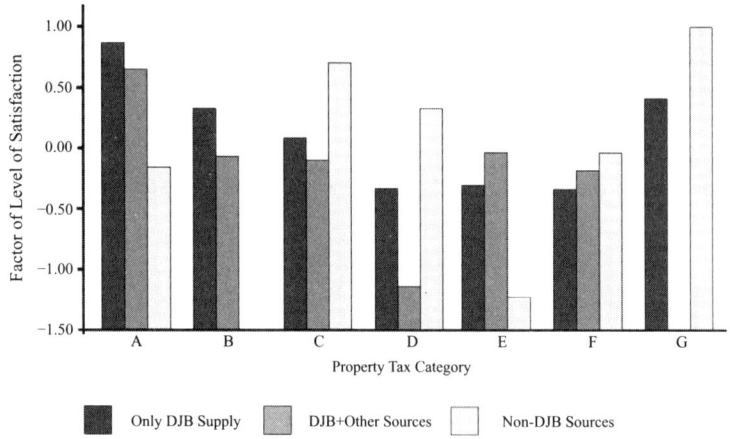

Source: Author.

Notes: 1. DJB sources mean the piped water supply.

2. DJB + other sources represent piped supply augmented with borewell.

3. Non-DJB sources of supply include only bore well and tanker in the absence of piped supply.

Combined Notes for Figures 7.8 and 7.9:

Factor of LoS is the combined score of three variables, namely quantity of water, regularity of supply, and duration of supply. Visual analysis of the data on the source of supply reveals three distinctive patterns in the mean value of the LoS regarding quantity of DJB water supply:

- LoS decreases from Strata I to Strata III and then increases to Strata IV. It also decrease from higher tax category of A to higher-middle category of D;
- LoS remains similar at higher-middle category of D to lower-middle category of F. Similarly, difference in the mean value of LoS in Strata II and III is marginal; and
- LoS increases at the lower tax category of G and at Strata IV.

Figure 7.10:
Estimated Mean of the Factor of LoS with Respect to Strata and Tax Categories

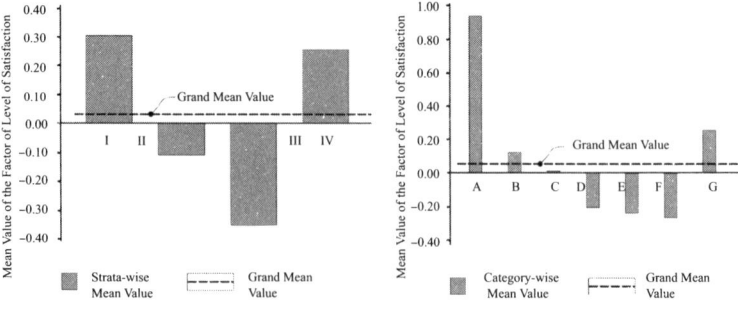

Source: Author.

Information on the LoS with the quality of supply has also been collected. There is a significant effect of strata on the LoS of the users with the quality of supplied water [$F(3,1094)$ = 13.484, p < 0.0001]. Estimated mean of the LoS of quality of water over the strata shows the expected patterns: higher values of LoS in Strata I and IV and lower ones in Strata II and III. Strata II and III have lower mean values than the grand mean (= 2.969). It again underlines the dissatisfaction of the people from middle and lower-middle level tax categories with water supply. Also, the lower strata of society, having very little expectation from the agencies, are content with whatever supply they get (Figure 7.11).

All these observations along with the data from the DJB reinforce the fact that the DJB supply is more in the higher socioeconomic groups and the LoS is also higher accordingly.

Figure 7.11:
Estimated Mean of User's LoS with the Quality of Water Supply

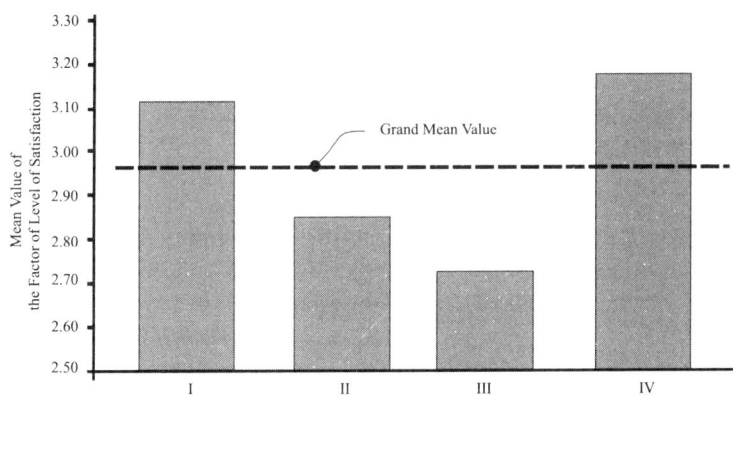

Source: Author.

Quantity of supply based on survey data also indicates two clear slabs, one with the top two strata and the other with the bottom two. However, despite lower DJB supply, lower socioeconomic groups have higher LoS. People from the lower socioeconomic strata, accustomed to the reality of inadequate service delivery of water supply, have low expectation from the state agencies. The satisfaction level of the people from the mid-strata of II and III (and the corresponding mid-level property tax categories, C, D, E, and F) is lower, underlining a generic phenomenon of dissatisfaction of the MIG with basic service delivery. The inequity across socioeconomic groups clearly shows that even within the colonies with *legal* property rights (including the *planned* colonies), the delivery of water supply has multiple gradations. It appears that supplying a scarce resource, like water, more to the richer sections of the society even amongst the so-called *legal* colonies is not a policy decision, but a governmentality technique. Such a technique somewhat reveals the strategic selectivity of the state in favor of the higher strata of the society and the spaces occupied by them in the city. Such activities of governmentality, Dean (1999: 17–23) mentions, result in "unpredictable consequences, effects, and outcomes."

It may be examined now whether multiple delivery situations across socioeconomic spaces are due to the correspondence and contradiction of physical territories or the scope of the jurisdiction of multiple agencies. I refer this condition as the overlapping urban governance.

Operational Zones of the DJB and Multiple Territories of Governance

In 2003, DJB reorganized its water distribution territory into 21 OZ for water distribution, sewage collection, and revenue collection within the zone, which is considered as a step to facilitate "better authority and accountability of distribution" (PWC et al. 2005: 13 [ch 2]). An earlier version of the same report, commissioned by the DJB, mentions that the "boundaries of these combined zones encompass complete wards of the civil administration, that is, the zone boundaries correspond to ward boundaries" and recommends the reorganization of distribution zones by making each zone distinct in terms of incoming water, distribution, maintenance and repair, NRW control, billing and revenue, etc. (PWC et al. 2004). The same document also highlights the need for arrangements to ensure an overall distribution system spatially coinciding with other factors (ibid.). The report's observation that the boundary of the DJB zone is co-terminus with that of the wards seems confusing, as the number of wards is more than that of the DJB zones. Unfortunately, what does not help in clarifying these doubts is the lack of availability of proper maps or drawings from the DJB, which, otherwise, would have been useful in an accurate delineation of these zones on a detailed map of Delhi (PWC et al. 2004).[20] There have been 134 wards before the delimitation in 2007–2008. One DJB OZ is likely to have within its physical territory an average of more than six MWs. In the pre-delimited situation, one AC, by and large, has contained two wards within its boundary. All these imply an inherent mismatch not only between different DJB zones, but also between DJB OZ and Assembly/ward boundaries.[21] In fact, the PWC report (PWC et al. 2005) proposes an extensive mapping of the engineering and revenue zones in order to delineate common boundary for both, for better accountability and services as well as for maintaining direct linkage between the services provided and charges for the services.

Interestingly, jurisdictional interrelationships of the Assembly and DJB zones are particularly important because each MLA under the "Special scheme for Grant-in-aid for development of sewerage and water supply in ACs" can suggest works for water supply and sewerage improvements to the extent of ₹50.00 lakhs in their constituency. In fact, an outlay of ₹3,500.00 lakhs, approved for the Grant-in-aid scheme in the Annual Plan in 2006–2007, was about 8 percent of the total budget for water supply (DJB 2006–2007; Table 7.5).

One would be curious to see the mutual interrelationships of the territory of jurisdictions and whether that has any relevance to water supply. Traces of injustice and inequity have already been observed in the delivery of water supply across various socioeconomic spaces, based on the property tax categorization of the MCD. Besides that, the planning zones of the DDA and the political jurisdictions may also have certain relationships with water supply across various socioeconomic spaces. The planning territory is important, because it is a policy framework. The MLALAD funding through the DJB also conveys the significance of the physical extents of a political territory, like the AC. Another territory of influencing the pattern of supply is the area covered by each UGR, which is referred here as the DJB service zones.

The objective, here, is to observe whether differences in water supply are affected by three territories of governance, namely the DJB service zones (representing the command area of UGR), Assembly-level political constituencies, and planning zones (popularly known as *planning subzones*). With the help of the DJB dataset, a set of factorial ANOVAs are conducted with FWS (as per the DJB data) as the dependent variable and territories of jurisdictions as independent variables (FWS*"Strata*DJB service zone*Assembly Constituency*Planning zone"). Bootstrap sampling method is used. The model is significant at 99 percent confidence level ($p < 0.001$). Among the dependent variables, strata has a significant effect on FWS [$F(3,18) = 3.669$, $p < 0.05$], so has the AC [$F(2,18) = 3.614$, $p < 0.05$]. The planning zone, again, has a partially significant effect on the FWS [$F(4,18) = 1.993$, $p = 0.139$] and the DJB service zones under the first level of the DJB UGR have no effect on the FWS at all [$F(1,18) = 0.145$, $p = 0.708$]. Estimated marginal means clearly show that the higher strata have higher mean value of the FWS (Table 7.11).

However, the best factorial ANOVA model to explain relationships among the majority of independent variables and their interactions is found by excluding the DJB service zone as an independent variable (FWS*"Strata*Assembly Constituency*Planning zone"). The model is significant at 99 percent confidence level ($p < 0.001$). Among the independent variables, strata has a significant effect on the FWS [$F(3,19) = 4.037$, $p < 0.05$]; so has the AC [$F(3, 19) = 13.997$, $p < 0.001$]; the planning zone has a partially significant effect on the FWS [$F(4,19) = 2.145$, $p = 0.115$].

The combined interaction of the AC and the planning zone has significant effect on the FWS [$F(1,19) = 5.021$, $p < 0.05$]. The strata in interaction with the AC has a partially significant effect on the FWS [$F(3,19) = 2.738$, $p = 0.072$] and similarly, the strata in combination

Table 7.11:
FWS as per DJB Data and Tax Strata: Relationship-II

Strata Based on Property Tax Category	Mean	Std. Error	95% Confidence Interval		Bootstrap for Mean[b]			
			Lower Bound	Upper Bound	Bias	Std. Error	95% Confidence Interval	
							Lower	Upper
I	.480[a]	.153	.158		.059	.114	.329	.790
II	.045[a]	.178	−.418	.328	−.102	.267	−.720	.331
III	−.283[a]	.167	−.633	.068	−.048	.220	−.773	.102
IV	−.640[a]	.251	−1.168	−.111	.048	.263	−.946	−.011
Grand Mean	−.001[a]	.089	−.188	.187	−.018	.102	−.219	.181

Source: Author.
Notes: Other independent factors: DJB service zone, AC, and planning zone.
[a]Based on modified population marginal mean.
[b]Unless otherwise noted, bootstrap results are based on 1,000 bootstrap samples.

with the planning zones, too, has partially significant effect on the FWS [$F(4,19) = 2.134$, $p = 0.116$]. Estimated marginal means clearly show that the higher strata have higher mean value of the FWS (Table 7.12). Pair-wise comparisons show that strata I has a significantly higher water supply component, that is, FWS, than other three strata ($p < 0.05$).

It is also worth observing at this point if the survey data also indicates any effect of the same territorial jurisdictions on water supply. Although each independent variable has a significant effect on the FDS ($p < 0.05$), it is essential to examine whether differences in the duration of water supply as per the survey data are affected by the combination of the factors mentioned before (independent variables), namely strata, DJB service zone, AC, and planning zone; data are subjected to ANOVAs with FDS as the dependent variable. The particular ANOVA model (FDS*"Strata*DJB Service Zones*Assembly Constituency*Planning Zones") best explains the combined effect of the factors on the duration of water supply all around the year.

Average duration of supply in the higher strata, I and II, is more than the grand mean, whereas that in the lower strata, III and IV, is less than the grand mean (Table 7.13). It once again confirms that there exist two broad slabs of water supply. The ANOVA model indicates that differences in the duration of water supply are affected by differences in the strata

Table 7.12:

FWS as per DJB Data and Tax Strata: Relationship-III

Strata Based on Property Tax Category	Mean	Std. Error	95% Confidence Interval		Bootstrap for Mean[b]			
			Lower Bound	Upper Bound	Bias	Std. Error	95% Confidence Interval	
							Lower	Upper
I	.548[a]	.149	.235	.860	.033	.102	.380	.781
II	−.045[a]	.174	−.408	.318	−.111	.276	−.740	.359
III	−.283[a]	.163	−.624	.058	−.032	.222	−.767	.088
IV	−.640[a]	.246	−1.154	−.125	.058[c]	.269[c]	−.943[c]	−.008[c]
Grand Mean	.004[a]	.087	−.179	.187	−.014	.099	−.217	.184

Source: Author.
Notes: Other independent factors: AC and planning zone.
[a]Based on modified population marginal mean.
[b]Unless otherwise noted, bootstrap results are based on 1,000 bootstrap samples.
[c]Based on 998 samples.

Table 7.13:
FDS as per Survey Data and Tax Strata: Relationship-I

Strata Based on Property Tax Category	Mean	Std. Error	95% Confidence Interval	
			Lower Bound	Upper Bound
I	−.037[a]	.036	−.108	.034
II	.479[a]	.064	.352	.605
III	−.560[a]	.043	−.645	−.475
IV	−.386[a]	.079	−.540	−.231
Grand Mean	−.051[a]	.027	−.104	.002

Source: Author.
Notes: Other independent factors: AC, DJB service zone, and planning zone.
[a]Based on modified population marginal mean.

$[F(3, 1041) = 17.997, p < 0.001]$, the ACs $[F(1, 1041) = 17.997, p < 0.001]$, and the planning zones $[F(10, 1041) = 60.905, p < 0.001]$. In this combined model, the effect of the DJB service zones is not found significant $[F(1, 1041) = 1.074, p = 0.30]$.

The model also shows that differences in the "factor of the LoS" regarding the quantity of water supply are affected by differences in the strata $[F(3, 1092) = 6.114, p < 0.001]$, the ACs $[F(1, 1092) = 26.854, p < 0.001]$, and the planning zones $[F(10, 1092) = 16.549, p < 0.001]$. In this combined model, the effect of DJB service zone is not found significant $[F(7, 1092) = 0.819, p = 0.336]$. The highest and the lowest strata, I and IV, have the average LoS more than the grand mean, whereas the average LoS in the middle strata, II and III, are less than the same (Table 7.14).

Table 7.14:
Factor of LoS as per Survey Data and Tax Strata: Relationship-II

Strata Based on Property Tax Category	Mean	Std. Error	95% Confidence Interval	
			Lower Bound	Upper Bound
I	.306[a]	.050	.209	.404
II	−.112[a]	.063	−.235	.011
III	−.351[a]	.058	−.464	−.237
IV	.257[a]	.110	.041	.474
Grand Mean	.021[a]	.032	−.042	.084

Source: Author.
Notes: Other independent factors: AC, DJB service zone, and planning zone.
[a]Based on modified population marginal mean.

Means and Opportunities toward Distributional Equity

Importantly, water supply, whether the FWS based on the DJB data, or both the FDS and the factor of LoS based on the survey data, seems to be least affected by the DJB service zone, when seen together with other territorial factors. Therefore, water supply may not be dependent much on the issue of coverage with respect to different DJB service zones within the study area for empirical observations. Discussions on uneven distribution at the broad territorial/geographical level, too, corroborate the same. Instead, the delivery of water supply appears to be more influenced by other factors, like the tax strata or the AC. The effect of planning zone per se is somewhat partially suggesting that areas situated in the different planning zones may not always have significant differences in water supply.

A couple of thoughts immediately come to mind: One, service zones for water supply may be reconstituted keeping in mind the limited influence of the planning and the existing DJB zones, and the other, the possibility of using the tax strata may be explored in addressing issues of inequity.

The reconfiguration of the service zone for water supply may also take into account political territories, like the AC, seeing its relevance in the empirical finding. The AC, interestingly, has significant effect on water supply. Also, within each constituency, the error variances are more or less found equal across groups of colonies, but significant inter-constituency variations are visible. These observations, indeed, bring out the importance of the political constituency (space) in water supply.

The property tax strata, if conceived somewhat like the "income charge" as mentioned earlier, may be used in tackling issues of vertical/distributional inequity. Consequently, aspects of affordability, pricing and the quantity of supply may be seen in that light. I will come back to this discussion a little while later.

The issue of vertical/distributional equity is lopsided at the moment, when the higher socioeconomic strata seem to enjoy higher water supply irrespective of the same norms of supply across all these strata. Various factorial ANOVA models clearly suggest that the strata or the socioeconomic spaces, even seen in conjunction with territorial factors of governance, have significant effect on water supply. Survey data, once again, reveals the existence of two broad slabs, one with Strata I and II and the other with Strata III and IV on the higher and lower half of the grand mean of the duration of supply. LoS highlights three dimensions. First, highest supply is availed by the highest socioeconomic groups (Strata I) having maximum LoS regarding the water supply. Second, despite lower DJB supply, higher LoS among the socioeconomic groups at the bottom indicates that people from the lowest strata of society are accustomed to the reality of inadequate service delivery of water supply. Hence, having a low expectation from the state agencies, they are seemingly *satisfied* with whatever supply they get. Third, the lower LoS in the mid-strata, that is, Strata II and III, somewhat, underlines a generic tendency of dissatisfaction of the MIG with the basic service delivery in the city.

The distributional inequity is found to be caused by the differential delivery across colonies sharing the supply network and it appears to be regardless of the locational proximity of the socioeconomic spaces of the so-called legal colonies, whereas the *illegal* (or *para-legal*) settlements, quite a few of them situated next to the legal colonies, have different supply norms and very often do not enjoy the same networks of supply. Hence, for legal colonies, the delivery varies and for the para-legal ones, the norms. Such readings bring back the notions of the *means* of supply and the opportunity of enhancing the *capability* to use such means. Less supply means less *means* and less *opportunity*.

Is it possible to address the issues of means and the opportunity in approaching vertical or distributional equity? Three aspects of dealing with such an intention, perhaps, would be affordability, pricing, and the quantity of supply.[22]

The question of affordability can be addressed by introducing the differential pricing system by charging more from the people who can afford more. The property tax strata may be useful in this regard. The prevailing situation is ironical to say the least: the "poverty premium" is to be borne by the urban poor for making alternative arrangement for the DJB supply. Preferably it should be the reverse; the way vegetables and other essentials cost more in the *high-end* colonies, per unit water, too, may cost more in colonies with higher property tax strata. To attend to the issue of accessibility to resources, the quantitative aspect of supply may now be brought to the fore. Interestingly, one can see the difference in the policy norms for per capita supply by the DJB and the CPHEEO/MoUD of the Government of India; the former suggests a benchmark of 121 lpcd across all settlements and the latter, a differential range of 70–200 lpcd on the basis of the legal status of colonies. Apparently, neither of the norms is delivered in its totality and it is difficult to notice any mechanism to ensure an equal level of minimum supply across all strata of society including the squatter settlements.

In fact, the findings suggest an uneven delivery pattern portraying the mentality of the state that is to supply a scarce resource, like water, more to the richer strata of the society. When the present sets of norms seem inadequate in providing the means of access and the opportunity to avail such means, perhaps, *a two-slab model* may be opted for, which, on the one hand, would guarantee a minimum quantity of water supply equally across all strata of the society and on the other, would introduce, depending on the affordability and the need of the strata, differential levels of the optimum supply.

Now, the question is: What are the methods to be adopted that bring us closer to the identification of the need and affordability? An attempt is made later in the concluding part of this work to address this question theoretically, where I turn to the notions of the "structure" and "agency," to underline the importance of the local knowledgeability (of the user groups) in recognizing the key issue of need and affordability.

Observations in the empirical research also suggest that within the city of Delhi exist conditions of differential delivery of the same policy norms as well as the provision of different norms. Policy norms are delivered at few places, whereas in most of the other places, they are not. Multiple conditions of living, then, create the *equity mosaic*, formed out of the heterogeneity of geographical and distributional conditions of inequities. I try to address such readings of urbanity from an understanding that the city is a site of heterotopias. I will return to these ideas later.

Overlaps of Political Spaces and Territories of Governance in Water Supply

Another dimension to the multiplicity may also be seen in the complexity out of the intersecting spatial jurisdictions and their mutual interactions. In particular, the political territories, like ACs, seem to have certain relationships with water supply. In that light, the possible reconfiguration of water supply territories coinciding with the political constituencies may be considered. This opens up two directions: one, concerning the broad relationships between the political spaces and the water governance and the other, regarding a more direct connection between the political responsibility and water supply. I will address the first point here and the second one later, while focusing on the political funding.

It is important to now look into the water governance-related aspects. Water governance is identified with the diverse spectrum of "political, organizational, and administrative processes" of the "development and management of water resources and delivery of water services" (Bakker 2003b: 5, Rogers and Hall 2003).[23] However the term is used here in a limited sense for the specific context of Delhi, a unique governance condition of the capital of the country and the city-state. Multiple political and administrative entities of governance have certain historical–spatial ambiguities in urban service delivery in Delhi and water supply is no exception to that.

The dual administrative existence of Delhi, the national capital and a city-state, has the three-tier political territories: PCs, ACs, and MWs (Table 7.15). As the capital of India, it has ministries, offices, residential colonies, and other institutions of the Government of India. Delhi's administrative structure was reformed from 1992 through an act leading to the formation of a separate Legislative Assembly and a Council of Ministers to govern the Union Territory of Delhi. Under this act, elections to the Delhi Legislative Assembly were held every five years since 1993. The act entrusted the Legislative Assembly with powers to govern the

Table 7.15:
Political Jurisdictions in Delhi

| | No. of Constituencies/Wards | | | No. of Seats | |
Political Jurisdictions	*Before Delimitation*	*After Delimitation*	*Election Year*	*INC*	*BJP*
Parliament of India:	07	–	2004	06	01
Lok Sabha	–	07	2009	07	00
Legislative Assembly	70	–	2003	47+	20+
of Delhi	–	70	2008	42+	23+
MCD	134	–	1997	37+	79+
	134	–	2002	108+	16+
	–	272	2007	67+	134+

Source: Compiled by author.
Notes: 1. *Lok Sabha* or the House of the People is the Lower House of the Parliament.
2. Reconstitution of constituencies has been done as per the Delimitation Act 2002.
3. INC: Indian National Congress; BJP: Bharatiya Janata Party.

state of Delhi, but these powers do not extend to law and order, land, and the police, which remain with the Government of India (Dewan 2004).[24] In addition to these two bodies, the MCD is the largest civic body to govern Delhi in terms of citizen-focused services (MCD 2005).[25] The MCD was established in 1958 under the Delhi Municipal Corporation (DMC) Act 1957 with the deliberative and the executive wings, headed by the Mayor and the Municipal Commissioner, respectively (Siddiqui et al. 2004).

It gets quite complicated when the land is controlled by the central government; several basic services to the (plots of) land are provided by the organizations controlled by the state government and the tax on the property, built on the land and serviced, is taken by the MCD. That is the concurrence and the conflict of jurisdictions in Delhi and the politics of differences out of it.

The Government of India has the final power over the MCD, whereas the GNCTD, the state government, has only limited controls on the corporation (Ghosh and Tawa Lama-Rewal 2005: 65, Siddiqui et al. 2004: 201). The MCD has gradually lost many of its tasks and responsibilities including water supply and sewerage which are now with the DJB under the state Government. The MCD also has no planning powers, which are with the DDA under the Union Government. There is a *lack of coordination*, in general, between all these agencies (Ghosh and Tawa Lama-Rewal 2005: 65).

A rather undefined relationship between the MCD and the GNCTD, often, suffers from conflicts of functional and political interests. One of the reasons was the enactment of the DMC Act, 1957 without an initial provision for a Legislative Assembly in Delhi, which has eventually been constituted in 1992–1993 (Jain 2009). The complex administrative triangle of the Central Government, the Delhi state Government, and the MCD remains somewhat unresolved because of the issues of land and financial approvals.

In fact, Delhi has a complex administrative set-up having about 120 public bodies involved in the management of the same territory of the national capital and the city-state (Jain 2009). Although for many services, primary responsibilities remain with key agencies: For example, the DDA is primarily responsible for urban planning, land management, and housing and the DJB, for water supply, etc., yet the overall governance for the provision of basic services continues to be a complex overlapping and sharing of institutional concerns and tasks (Jain 2009, Pinto 2000, Singh and Shukla 2005; Figure 3.1).

The relationship between the MCD and the GNCTD is particularly important in water governance. Historically, MCD had the control on water supply which was later entrusted with the GNCTD (Table 7.16). The administrative structure of the MCD, namely the "Weak Mayor Council System with Commissioner led Administration" initiated in Mumbai and then followed in Delhi and Chennai, is repeatedly identified with the bureaucratic inefficiency and "conflicts when the Commissioner representing state government of a party different than that is in power in Municipal Corporation" (Nallathiga 2008, Pinto 2000).[26] The importance of political parties in the process of governance is also of certain significance,

Table 7.16:
Institutional Responsibilities of Water Supply over the Years

Organization	*Period of Operation*	*Judicial Instrument*	*Under the Jurisdictions of*	*Functions and Responsibilities in Terms of Water Supply*
DJB	1998 onwards	Delhi Water Board Act, 1998	GNCTD	Procurement of raw water, treatment, and supply to MCD, NDMC, and DCB and responsible for distribution of water supply and related works in MCD area
The Delhi Water Supply and Sewage Disposal Undertaking (DWS and SDU)	1958–1998	MCD Act, 1958 by the Parliament	MCD	Procure, treat, and transmit water for the entire Union Territory of Delhi and to distribute water in the MCD area
The Delhi Joint Water and Sewerage Board (DJWS and SDU)	1926–1958		Not Available	Operation and maintenance of Delhi Water Works and supplying water in bulk to the five local bodies in Delhi

Source: Compiled by author from: IL&FS EcoSmart Ltd (2006).

which is addressed in the case study selection for this work. From an account of deliberations in the Delhi Assembly, its differences with the DDA and the MCD become quite obvious (Tawa Lama-Rewal 2005).[27]

Then, a sense of curiosity develops about the involvement of Delhi Assembly in water-related issues, which is also connected with the MLALAD funding through the DJB. In that direction, one can decipher three broad trends from the study conducted by Stephanie Tawa-Lama Rewal in 2005 on the issue of the urban governance in Delhi:[28]

- Water-related issues remained under low priority in terms of the legislative proposals put up by the ministers in the first and the second *Vidhan Sabhas* (Assembly);
- MLAs seemed to be more concerned with the water-related issues and raised more Private Member's Bill/Resolutions and short discussions, which were higher in number in the first *Vidhan Sabha* than in the second. But, despite higher occasions of short discussions on water-related issues, it was sixth in the priority list; and
- It appeared that MLAs were more worried about unauthorized colonies than about the slum development. That might have happened because of their own interest to use the MLALAD fund for bringing basic facilities to such colonies.

Some of the basic observations, emerging out of the discussions on the issues related to overlapping governance in the provision of water supply, may be summed up as the following:

- The land is with the central government (DDA), water supply on that land is with the state government (DJB), the tax on the property (or houses) to which water is being supplied is collected by the local body (MCD).
- The MCD takes varied property taxes without any control on water, hence, cannot provide better facilities regarding water supply for the people who pay tax.
- Different political parties holding different governments, often, create political and operational conflicts.
- The MLALAD seems to indicate the interests of the MLAs in ensuring water supply in their constituencies, especially in the unauthorized colonies.

It leads me to the next discussion to find out how political funding plays a role in the operation and maintenance of water supply.

Notes

1. Anthropologist Escobar (n.d.) defines in his website the notion of "political ecology as the study of economic, ecological, and cultural distribution conflicts. These conflicts arise out of economic, ecological, and cultural difference." Elsewhere, Escobar (1999) identifies that political ecology observes multiple interrelated biophysical and historical practices and the key objective of political ecology is "to understand and participate in the ensemble of forces linking social change, environment and development." Instead of conceptualizing politics as the end-result, Swyngedouw (2009), on the contrary, attempts to bring back the argument of politics to the center of the political ecology discourses.
2. Delhi Joint Water and Sewage Board came up for the first time in 1926.
3. The infrastructure for water supply and sewerage collection in these territories is owned by those authorities and, consequently, is not the responsibility of the DJB. The same applies to sewerage as well. Wastewater of these areas, collected by the NDMC and DCB, is received by the DJB at a number of locations and subsequently conveyed and treated, and the DJB is not responsible for storm water collection and drainage (PWC et al. 2004).
4. Water from the Yamuna River is abstracted both directly from the river at Delhi and indirectly via the Western Yamuna Canal, which also delivers water to Delhi. Raw water from the Bhakra canal and the Ganga River is conveyed via the Upper Ganga and Bhakra (Narwana Branch) canals, respectively. The water allocations for the DJB from these sources are: 1,835 TCMD (404 MGD) from the Upper Yamuna River Board, 1,213 TCMD (267 MGD) from the Bhakra–Beas Management Board, and 1,223 TCMD (269 MGD) from the Ganga River. A total of 4,271 TCMD (940 MGD) theoretically available water translates into an actual availability of 3,726 TCMD (820 MGD) due to losses from the canal systems and capacity limitation in the raw water transmission to one water treatment plant from the Western Yamuna Canal.

   The mentioned information is based on PWC et al. (2004).
5. *An Appraisal of Annual Plan 2004–05*, prepared by the planning department of Government of NCT of Delhi, mentions:

   > [The] JJ [*Jhuggi Jhompri*] clusters had been provided 106 MG water through tankers and the distribution network was improved with the

installation of 3 new tube wells, 60 DBHP, laying of 4.38 km of new water mains and replacement of 3.03 km. of old water mains. (Government of NCT Delhi 2004–05: 110–112)

6. Total domestic water demands for 2004, 2005, and 2006 were 2,076, 2,143, and 2,210 (all in TCMD), respectively, and the data follow a linear regression $67x + 1808$. Thus, in the absence of the water demand data for 2001, the regression equation leads to a demand figure of 1,808, 1,875, 1,942, and 2,009 (all in TCMD) for 2000, 2001, 2002, and 2003, respectively. Data source: PWC et al. (2004); DDA (2007a, 1990).

7. Such differences in the calculation of water demand have been continuing for some time. Earlier, in two occasions, DJB anticipated the water demands for 2006 and 2011 with 270 lpcd assumptions, whereas DDA's projections for the same time-frame were 360 lpcd (DDA 2007a: 150).

8. An analysis of the PWC report (2004) by the NGO Parivartan (ca. 2005) makes the following observations:

> DJB targets to reduce NRW from 55 percent to 34 percent in 3 years, whereas internationally, companies failed to reduce NRW. In Philippines, Manila Water Company (one of the companies short-listed for Delhi) promised to reduce NRW to 16 percent by 2001, but was still losing almost 50 percent water by that time. In Puerto Rico, the Rican Office of the Controller estimated that six years after handing over, Puerto Rico was still losing almost 50 percent of its water through leaking pipelines.

9. These "flash points" include "up-market areas," such as Defence Colony, Vasant Kunj, Greater Kailash, and "middle-class localities," such as Patel Nagar, Malviya Nagar, Dwarka, Kalkaji, Tilak Nagar, Vikas Puri, and Shahpurjat, and the "LIG localities," such as Sangam Vihar, Narela, Najafgarh, Uttam Nagar, and Dabri (Navdanya 2005: 34).

10. A classification of informal settlements in Delhi:

- JJ clusters, the slum clusters, or squatter settlements, which, as per survey in 1994 by the S&JJ Department of MCD, were about 1,080 in numbers accommodating 4.8 lakhs households.
- Resettlement colonies, which were about 46 in numbers, developed mainly on the outskirts of the city to resettle about 2.16 lakhs squatter families in a highly subsidized plot of approximately 18 sq. meter of land. These colonies face inadequate infrastructure such as water supply, sewerage, drainage, garbage disposal, electricity, schools, hospitals, roads, etc. The MPD 2021 suggests the "in-situ rehabilitation" option as an alternative strategy.
- Unauthorized/regularized colonies, which came up on *private* land in violation of the Master Plan or zonal plan regulations. 567 out of 607

listed such colonies were regularized till October 1993, but many more of those have reportedly come up since then (DDA 2007a: 21).

• Legally notified slum areas, declared/notified as slum areas under section 3 of the Slum Areas (Improvement and Clearances) Act, 1956, where buildings were found "unfit for human habitation" for lack of safety, health, or morals for living. Majority of such notified slums has been identified in the walled city of Shahjahanabad and its extensions.

• Pavement dwellers estimated about 70 thousand people, living on the pavements in busy market places in the city where they work as wage earners.

• Urban villages, about 106 villages on the outskirts of Delhi, which, initially were left out loosely by the MPDs, became urbanized in a haphazard and unplanned manner. These settlements were initially outside the jurisdiction of MCD and as a result, continue to face problems in the facilities like, assured potable water, surface drainage system, and sanitation arrangement.

Sources: Dhar Chakrabarti (2001) and DDA (2007a).

11. In the report, titled *Water Supply Schemes Under Water Bulk for the Year 2004–05* prepared by the Delhi Jal Board, details of schemes planned under "Loan Works—RSC, urban villages and unauthorized colonies for the year 2004–05" included for water supply to urban villages an amount of ₹300 lakhs, to resettlement colonies, ₹400 lakhs, and to unauthorized colonies, ₹3,400 lakhs (DJB 2004–2005: 91–104). Under the "Grant-in-aid JJ cluster" for the improvement of water supply and sewerage facilities in JJ clusters, an amount of ₹750 lakhs was also allotted (ibid.).

12. Survey conducted in Sanjay Colony near Okhla in South Delhi by Dimri and Sharma (2006) observes the following:

• Average expenditure on water by the residents is ₹236.00 per month per household and the average household water consumption in a month comes to 5,280 liters.

• For a family with five members, available quantity of water is about 35.2 lpcd.

• Average cost of consumption of water, here, turns out to be around ₹45.00 per kiloliter as compared to ₹6.67– 11.83 per kiloliter paid by the consumers across the city. That amounts to a "poverty premium" of 4–7 times on water.

• Furthermore, the quality of water they consume is inferior to that supplied by the DJB.

The work of Dimri and Sharma (2006) builds up on the notion of "poverty penalty" coined by C.K. Prahalad in his book, *The Fortune at the Bottom of the Pyramid* (2005), in which he compares the premium paid by residents of Dharavi, Asia's biggest slum, with Warden Road, one of the popularly known posh areas of Mumbai.

13. Population data is based on colony-wise information from DJB sources.

14. Total quantity of water supply (in lpcd) varies marginally from the filtered water supply as in some colonies other sources like, bore well and tankers are used for supplementing the DJB supply.

15. All inter-variable correlations are significant at 0.01 level ($p < 0.00001$). None of the correlation coefficients are more than 0.9 thereby negating a possibility of multicollinearity. Also, the determinant (= 0.047) is greater than 0.00001. KMO statistic (normally varies between 0 and 1) of a value (=0.768) is good and suggests that factor analysis is appropriate for these data. Here, the significant test tells us that the R matrix is not an identity matrix; hence, there are some relationships between the variables I hope to include in the analysis. For this data, Bartlett test is also highly significant ($p < 0.0001$), and therefore, factor analysis is appropriate.

16. Only the eigen value of component 1 (eigen value = 3.105) is more than 1. Average communalities of the components is 0.776 which is greater than 0.6. Also, the Scree plot shows that the graph flattens after component 2. Hence, component 1 is extracted here. The eigenvalues associated with each factor represent the variance explained by that particular linear component. First factor explains about 77.62 percent of the cumulative variances, which is very high. Principal component analysis works on initial assumptions that all variances are equal; hence, before extraction all are 1. Communalities after extraction reflect the common variances in the data structure, e.g., 87.2 percent of the variances associated with factor 1 are common, or shared variance. Another way to look at these communalities is in terms of the proportion of variances explained by the underlying factors.

17. The F test indicates the effect of property tax strata. This test is based on the linearly independent pair-wise comparisons among the estimated marginal means.

18. Apparently very high percentage of households (about 87 percent in summer and 92 percent for the rest of the year) gets water every day. Average frequency of supply shows marginal variation between supply in summer (6.61 days/week) and that in the rest of the year (6.74 days/week). Strata-wise variation across the year including summer months is also nominal ranging between 6.2 and 6.8 (days/week).

19. Both the variables on the duration of supply, namely the duration of supply in summer months (May–July) and the duration of supply during the rest of the year, have been corrected for normality by transforming them into square roots. Here, the intention is to understand relative variations in these variables and not to predict absolute values in particular. For such a purpose, this is a widely accepted method.

20. With reference to DJB maps, requested by the Legislative Assembly, M.M. Mahto, Member (Administration), vide letter (F. no. 01/66/05-AC(w)/3205),

dated December 18, 2008, wrote to the Principal Secretary (Home), Government of NCT of Delhi, on the subject "Classification of maps pertaining to water and sewer networks in Delhi Jal Board." He communicated that the DJB maps related to important installations are "confidential documents" and expressed the reservations of the agency (the DJB) regarding the sharing of maps due to "security risks".

21. The final report by PWC et al. (2005: 1–2 [ch 2]) observes:

> There are multiple units responsible for ensuring service delivery to customer, which causes splitting of responsibility and accountability. For example, in case of water distribution, the responsibility of civil maintenance is with distribution zones, and O&M (Operation and Management) of Booster Pumping Stations (BPS) in a zone are the responsibility of Electrical & Mechanical (E&M) Divisions.
> There is a mismatch of zonal boundaries, which complicates the task of compiling complete information of a zone/geographical area. In the absence of zonal information, it becomes difficult to identify O&M issues to be addressed and prioritization of corrective actions. It also inhibits identifying single point of authority for improving service delivery in a zone/geographical area.

22. I have already discussed earlier in this book, how property tax can be conceived somewhat like Harvey's formulations of "income charge" that considers a combined effect of earning, benefits, resource availability, and the price of resources (Harvey 1975[1973]: 53–54; also refer to Figure 4.2).

23. Karren Bakker in *Good Governance in Restructuring Water Supply: A Handbook*, prepared for the Federations of Canadian Municipalities, mentions the following.

> The term "governance" refers, in general, to the relationship (economic, social and political) between a society and its government, or between an organization and its governing entity. Governance is often referred to as the "art of steering societies and organizations." Specific definitions of governance vary depending on context, and the term can be used in different ways. Governance, according to this definition, includes (but is broader than) formal structures of government. (Bakker 2003b: 5):

> Bakker adopts the above notion from the work of Rogers and Hall (2003), titled *Effective Water Governance*.

24. At this point of time, matters covered by entries 1, 2, and 18 of the state list of the 17th Schedule; public order, police, and land have been kept outside the purview of the Delhi Legislative Assembly (Dewan 2004).
25. MCD mentions in its *Civic Guide* that despite other two civic bodies, NDMC and Delhi Cantonment, the MCD has in its jurisdictions about 98 percent of the total area of Delhi holding about 94 percent of the total population of the city-state (MCD 2005: 7–8)
26. Local governance in the MCD is under two divisions of functions: the deliberative wing of elected councilors for policy and regulation functions and the executive wing of the commissioner, additional commissioners, and the heads of departments for the administration and executive powers (Jain 2009, MCD 2005: 15–19, Shah and Bakore 2006: 27, Tawa Lama-Rewal 2005). Shah and Bakore (2006: 27) mention that:

> In Delhi, until 2007, there were on an average 100000 people in each of the 134 municipal wards, and Wards Committees covered an average population of more than one million people. But a re-delimitation of wards took place prior to municipal elections in April 2007, making the average population of the new 272 wards closer to 50000 people—still a large figure by any standard.

27. The following findings may indicate such disagreement (Tawa Lama-Rewal 2005):

   - DDA was the direct subject of 7.9 percent of short duration discussions in the First *Vidhan Sabha* and 9.4 percent of that in the Second *Vidhan Sabha*.
   - 11.3 percent of such discussions were also devoted to complaints against the work of the MCD in the Second *Vidhan Sabha*.
   - Private member bills/Resolutions and short discussions on housing issues, for example 11.4 percent of Private member bills/Resolutions in the First *Vidhan Sabha*, also pointed at "recurring conflicts between MLAs and the Delhi Development Authority" as housing is under the DDA.

28. Stephanie Tawa-Lama Rewal conducted a study in 2005 to find out how, in the context of Delhi, health services can indicate the state of urban governance. Tawa-Lama Rewal (2005) noted that although Delhi's Assembly (*Vidhan Sabha*) did not meet too often. In a year, on an average, 26 sittings happened in the first *Vidhan Sabha* and 16 in the second. Based on her study, I observe following facts on the Assembly's interest on water-related issues:

   - One out of 44 and one out of 39 of major areas of legislative proposals in the first and in the second *Vidhan Sabhas*, respectively, were related to water. It indicates low priority on water-related issues in the legislative proposals put up by the ministers.

- In the first *Vidhan Sabha*, 6.3 percent of the Private Member's bill/resolutions and 7.5 percent of the short discussions of the MLAs were devoted to the water sector, which, in terms of priorities, were the fifth and the second. In the second *Vidhan Sabha*, water-related issues were the object of only 1.8 percent of the Private Member's bill/resolutions. Despite 7.5 percent of short discussions on water-related issues, it was sixth in the priority list.
- MLAs' concerns for water issues in unauthorized colonies (11.4 percent of Private member bills/resolutions in the first *Vidhan Sabha*) were found more than those for slum development (3.8 percent).

8

Water Supply, Political Funding, and Socioeconomic Spaces

The central question, here, is: how does political patronage relate to the socioeconomic spaces (indicated by the property tax strata) in Delhi? There are three levels of political funding schemes corresponding with three political territories, namely MPLAD scheme, MLALAD scheme, and the MCD Councilor Fund Scheme.

The MPLAD scheme was introduced on December 23, 1993. The MPLAD scheme is presently under the Ministry of Statistics and Programme Implementation, which is responsible for formulating the policy, releasing funds, and prescribing the mechanism to monitor the implementation of the scheme. The objective of the scheme is to enable MPs to recommend works of developmental nature to be taken up in their constituencies "for creation of durable community assets and for provision of basic facilities including community infrastructure, based on locally felt needs" (Ministry of Statistics and Programme Implementation 2010–2011: 3). Since the inception of the scheme, certain infrastructure provisions such as drinking water, primary education, public health, sanitation, roads, etc., have been the priorities (ibid.: 4). The scheme, launched in 1993–1994 with ₹5 lakhs per annum allotment per MP, has increased the amount over the years. At present, each MP can spend to ₹5 crores per annum under the scheme.

In 1997, MCD, too, initiated an allocation of ₹35 lakhs for the ward-level development under the Councilor's Fund, which over the years increased to ₹2 crores in 2009. The scheme is supposed to work as an incentive for the councilors to undertake development projects in their wards. In general, the fund utilization in terms of spending the allocated sum of money in the wards has not been found up to the mark.

The MPLAD spending on projects through the DJB is few and far between, and the Councilor Fund in Delhi cannot be spent through the DJB, an agency under the Delhi state government. The MLALAD, then, remains as the only relevant political funding for water supply in Delhi.

Each MLA under the "Special scheme for Grant-in-aid for Development of Sewerage and Water Supply in ACs" can suggest works for water supply and sewerage to the extent of ₹50.00 lakhs in the respective constituency. An outlay of ₹3,500.00 lakh was approved for the Annual Plan 2006–2007 which was about 8 percent of the total budget for operation and maintenance of the water supply and, interestingly, 1.25 times the expenditure on water supply to regularized/unauthorized colonies, JJ clusters, urban villages, and resettlement colonies together (DJB 2006–2007; Table 7.5). Specifically, in a five year period (2004–2009), ₹1,618.00 lakhs under the MLALAD have been used within the territory of the South II and III of the DJB OZ, which is the empirical context of this work. Hence, it is important to observe the overall spending of the MLALAD to confirm the pattern and the effect of the socioeconomic spaces on it.

Following the union government, the state government of Delhi initiated in September 1994, the MLALAD scheme for the "Strengthening and Augmentation of Infrastructure Facilities in Assembly Constituencies," which enables each MLA of Delhi to recommend small developmental works (of capital nature not exceeding ₹70 lakhs each) in his/her constituency through the allocated funds of ₹2 crores per annum (Department of Urban Development 2009). In 2011, following the demand by the MLAs, the fund was increased to ₹4 crores per annum. The Urban Development Department of the Government of Delhi is the node responsible for releasing funds and overall monitoring of this scheme. DJB is one of the agencies through which works could be executed (Table 8.1).

The total amount allocated for each MLA is divided into two components, namely the mandatory component of ₹1 crore to be directly disbursed by the Department of Urban Development to MCD and NDMC for taking up of works on the basis of written requests to them by each MLA and the remaining discretionary amount to be released by the same department for works to be undertaken by any of the above designated agencies at the insistence of the MLA (Department of Urban Development 2009). All the implementing agencies are required to submit utilization certificates to the Urban Development Department by the end of the financial year or on completion of work, whichever is earlier.

Table 8.1:
Agency-wise Expenditure of the MLALAD Funds (in lakhs)

Agency	Funds Released for 2002–2003	% of Funds Released for 2002–2003	Unspent Balance of Earlier Years	Total Fund Available	Expenditure	Unspent Balance	Expenditure as % of Total Fund Available
MCD	11,919.27	89.62	6,778.69	18,697.96	9,566.14	9,131.82	51.16
NDMC	435.00	3.27	261.46	696.46	425.92	270.54	61.15
DJB	158.23	1.19	246.96	405.19	242.71	162.48	59.90
I&FC	326.90	2.46	72.94	399.84	212.08	187.76	53.04
DDA	47.50	0.36	–	47.50	–	47.5	0.0
PWD	4.70	0.04	–	4.70	–	4.70	0.0
S&JJ	65.11	0.49	–	65.11	–	65.11	0.0
DVB	4.23	0.03	–	343.29	4.23	339.06	1.23
Discom	339.063	2.55	–	–	–	–	–
Total	13,300.00	100.00	7,360.05	20,660.05	10,451.08	10,208.97	50.59

Source: Compiled by author from Government of NCT of Delhi (2003).

Notes: 1. Tentative progress report for the year 2002–2003 as on March 31, 2003.
2. MCD: Municipal Corporation of Delhi, NDMC: New Delhi Municipal Council, DJB: Delhi Jal Board, I&FC: Irrigation and Flood Control, DDA: Delhi Development Authority, PWD: Public and Works Department, S&JJ: Slums and *Jhuggi Jhopri*, DVB Discoms: Delhi Vidyut Board Distribution Companies.

Following are some of the generic observations on spending of the MLALAD fund:

There seem to be the occurrence of "storming," that is, the increase in spending in the final years of the five-year-plan period (Sruthijith 2003: 295–303). It appears that MLAs, as a rule, use most of the fund in the last two years and the least in the first year after their election. It is difficult to decipher whether this is due to either the delay on the part of the MCD and other civic agencies or the lack of initiatives from the MLAs. Perhaps, such pattern of spending makes the work most visible and would last in the memory of the electorate during the next election. A newspaper report suggests that only 17 percent of the MLAs could use their funds in the financial year 2012–2013 (Rahman 2013).

There have been instances, in general, of the inability of the executing agency to spend the funds. In 2002–2003, the MCD could only spend about 50 percent of the fund at its disposal from the MLALAD (Table 8.2). However, the expenditure of the MLALAD funds has been increased over the years; for example, 3.6 percent of the allocated funds were spent during 1999–2000 which rose to 52.2 percent in 2002–2003 (Table 8.2). Similar observations can also be made with respect to the spending of the MPLAD funds and the councilor fund; for example, in 2008–2009, 188 councilors (out of 272) had spent less than ₹1 crore, 82 had spent less than ₹50 lakhs, and 27 councilors had spent less than even ₹30 lakhs (Sarkar 2009).

The spending of funds for the purpose not recommended in the guidelines is quite common, especially in the beginning. Instead of addressing the original purpose of the asset-creation, 67–82 percent of the works in the MLALAD, as mentioned in the CAG report of 2006, were primarily used for repair and maintenance (Express News Service 2006 and PAC 2005).

Table 8.2:
Consolidated Statement of MLALAD Funds (in lakhs)

Year	Total Balance/ Release	Expenditure Incurred	Unspent Balance	Expenditure as % of Total Balance
1999–2000	776.78	28.22	748.56	3.63
2000–2001	2,032.11	381.24	1,650.87	18.76
2001–2002	3,969.80	2,914.95	1,054.85	73.42
2002–2003	11,956.09	6,241.73	5,714.36	52.20
Total	18,734.78	9,566.14	9,168.64	48.93

Source: Compiled by author from Government of NCT Delhi (2003).

Political differences were caused due to the Congress-led Government at the state and the BJP-led MCD (Sarkar 2009). One of the recent flashpoints is the state government's suggestion of the reduction in the councilor fund. There have been recurring tensions of governance between the preexisting MCD and the newly formed Delhi state government since both administer roughly the same area (Kundu 2006, Siddiqui et al. 2004). Also, there has been a perception of the MCD being poorly managed. This gap was further widened by the proposal, supported by the Delhi state government to decentralize the MCD and to establish a separate agency for the slum and urban poverty issues. The proposal was, nevertheless, unanimously rejected by the MCD councilors. On the contrary, one of the popular demands of the MCD has been to reintegrate with itself the departments of water supply and sewer maintenance and electricity, which are presently under the para-statal agencies, like the DJB and the Delhi Vidyut Board, respectively (Kundu 2006, Siddiqui et al. 2004).

In the empirical analysis, I would like to address following key issues and concerns raised so far. First, is there any *patron–client* relationship evident from the spending of the political funding? Scott (1972), Chakrabarty (1989), and Chatterjee (2004) suggest possibilities of the "political patronage" for poorer sections of the society. In such a situation, political funds are expected to be spent more on the lower strata of socioeconomic groups and spaces (colonies). This is connected with the idea of the *vote bank* politics as well, when political representatives try to soothe their supporters by spending more in those colonies. It is quite understandable that the socioeconomic spaces of the lower strata of society tend to accommodate more people than the higher ones and any spending in those areas would literally reach more people.

Second, does the state favor the rich? In contradiction to the first question, the underlying assumption here extends the argument of the "selectivity of the state" put forward by Jessop and others (Brenner 2004, Jessop 1990, Jessop et al. 2008, Park 2008). It is also connected with Harvey's (1975[1973]) notions of the political patronage of the influential and the rich as well as the "growth coalition" construct by Molotch and others (Lasswell 1936, Logan and Molotch 1987[2002], Logan and Swanstrom 1990[2005], Mollenkopf 1992, Molotch 1976, 1993).

And finally, is there any *storming* of expenditure? The trend to spend more funds toward the election year seems obvious for enhancing the visibility and the *shelf-life* for fetching votes for the next election.

What these three questions try to raise is quite close to Agamben's argument (2005[2003]) of the "state of exception". To sum up, what these question raise is whether the spending of the political funds favors the spaces occupied by either the poorer or the richer socioeconomic strata, or are the political funds selectively spent around the time of the election.

I make an attempt to address, yet may not answer this particularly complex question through the empirical observations on two sets of data: one, the MLALAD spending through the DJB (MLALAD and MLA Priority Fund allocated to DJB over the years) and, the other, overall MLALAD spending on other sectors through MCD (MLALAD Combined Fund Spending over the years). The fund spent through the MCD does not include water supply-related funds. Since maximum share of the fund is usually spent through the MCD, for example, the spending of about 90 percent of the total fund in the year 2002–2003, the analysis of the second dataset would give an overall pattern of the spending of the fund in comparison with the utilization of the fund through the DJB in the first dataset (Table 8.1).

The hypothesis is: Multiple techniques of governmentality influence practices of political patronage reflected in the spending of the political funds that overwhelmingly favor neither the higher nor the lower socioeconomic strata, yet have tendencies to favor spaces occupied by richer sections of the society.

The Pattern of Spending of the MLALAD Fund Allocated to DJB

The major political funding for the DJB is the MLALAD of about 50 lakhs per year per MLA. The dataset on the MLALAD and MLA Priority Funds allocated to the DJB is compiled from the information available at the DJB website, for the period of six years (2004–2009).[1] The dataset includes the work only from South II and III zones of the DJB where the study area, discussed in the chapter 7 on water service delivery, is also located. Within these two zones, funds for seven ACs, namely Dr Ambedkar Nagar (AC 34), Hauz Khas (AC 9), Jangpura (AC 5), Kalakji (AC 7), Kasturba Nagar (AC 4), Malviya Nagar (AC 8), Okhla (AC 6), have been utilized.

The data is analyzed through two sets of information: the first, the "amount allocated" to the DJB and the second, the "amount allocated per capita," obtained by dividing the "amount allocated" by the population of the colonies available from the DJB sources. Within the period of 2004–2009, total MLALAD "amount allocated" has been ₹1,178.74

lakhs and the allocation, in other words, the redistribution of the fund is found higher in the lower half of the socioeconomic strata (Strata III and IV) and lower in the upper half (Strata I and II), almost 68 percent and 32 percent, respectively (Figure 8.1).

Figure 8.1:
Summary of Spending of the MLALAD Fund through DJB (2004–2009)

A. Strata-wise allocation of funds (Amount in ₹ in lakhs)

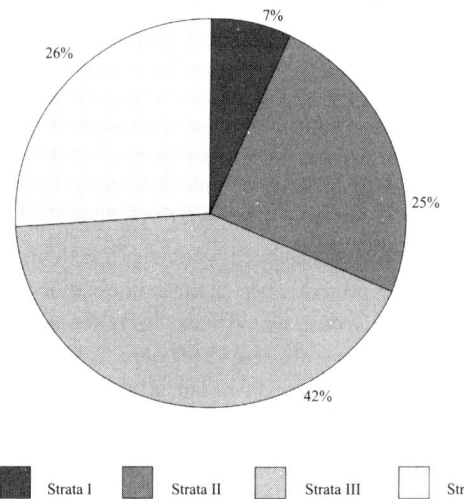

B. Average Amount Allocated per capita

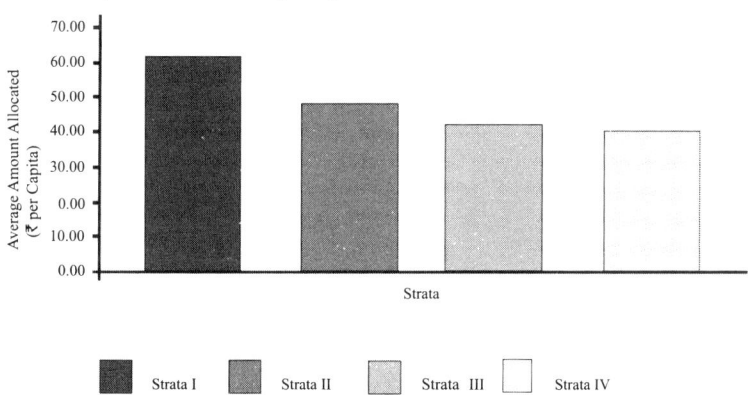

Source: Author.
Note: The dataset includes the work from South II and III zones of the DJB where the study area, discussed in the Chapter 7 on water service delivery, is also located.

In fact, two-thirds of the fund has been allocated to the colonies occupied by the lower half of the socioeconomic spectrum (Figure 8.1). Spending more on the projects to benefit the larger section of people may, perhaps, be expected in a patron–client relationship that the politicians have had with their electorate.

But looking at the per capita spending, one may gather a different understanding. The spending of the fund would actually benefit the higher strata more. Average per capita allocation of the MLALAD fund is found higher at spaces occupied by the higher strata of society (Table 8.3, Figure 8.1). Per capita allocation has a significant positive correlation with the allocation of funds as well ($p < 0.01$). It indicates that the people from the higher strata, in effect, have got more money per head.

There has been a tendency to spend smaller amount per project and, in turn, cover more number of colonies. This is a typical trait of the politicians to please wider sections of the population. The amount allocated for around 75 percent of projects has been less than ₹8.50 lakhs, whereas for about 80 percent of the projects, per capita money allocated was about ₹64 only. Average project cost would have been about ₹6.0 lakhs and the per capita allocation was about ₹45.00 only. ANOVA results show significant effect of strata on the "amount allocated" (in lakhs) in the MLALAD through the DJB [$F(3,189) = 3.612$, $p < 0.05$)].[2] Hence, the variation of overall MLALAD funding through DJB is found significant across strata.[3]

Among the ACs, maximum amount (₹316.97 lakhs), was allocated at Dr Ambedkar Nagar (AC 34), whereas average per capita (MLALAD)

Table 8.3:
Strata-wise Allocation of the MLALAD Funds through DJB (2004–2009)

Property Tax Strata	Amount Allocated per capita (₹)			Amount Allocated (in lakhs)	
	Mean	*N*	*Std. Deviation*	*Sum*	*N*
I	61.58	12	48.23	81.81	18
II	48.08	22	53.58	291.45	50
III	42.13	53	53.84	500.79	95
IV	40.41	18	39.34	304.69	34
Total	45.30	105	50.69	1,178.74	197

Source: Author.

amount allocated (₹112.09) is the highest at Malviya Nagar (AC 8) (Table 8.4). Higher allocation of funds in the constituency of Dr Ambedkar Nagar might have been because of the presence of a large number of lower socioeconomic groups. ANOVA results show that there are significant effects of the AC (independent variable) on the "amount allocated" (in lakhs) [$F(6, 186) = 4.332$, $p < 0.0001$] and "amount allocated per capita" (₹ [$F(6,94) = 10.042$, $p < 0.0001$]. It means political jurisdictions, in this case the ACs, have significant effect on both the total funding and the per capita funding of the MLALAD through DJB.

Pattern of Spending of the MLALAD Fund Allocated to the MCD

The MLALAD fund spent through the MCD has been observed within a time period from 2004–2005 to 2007–2008 across 78 colonies in four ACs, namely Hauz Khas (AC 9), Kalkaji (AC 7), Malviya Nagar (AC 8), and Okhla (AC 6), in which the study area, discussed in the Chapter 7 on water service delivery, is also located.

Two sets of information are available in this particular dataset: One, the "cost of work" and the other "up to date expenditure" indicating

Table 8.4:
Assembly-wise Allocation of the MLALAD Funds through DJB (2004–2009)

Old Assembly (No.)	Amount Allocated per Capita (₹)			Amount Allocated (in lakhs)	
	Mean	N	Std. Deviation	Sum	N
Dr Ambedkar Nagar (AC 34)	36.81	23	45.50	316.97	33
Kasturba Nagar (AC 4)	35.32	13	21.39	66.94	14
Jangpura (AC 5)	27.06	31	21.06	234.51	47
Okhla (AC 6)	20.29	8	32.12	191.04	33
Kalkaji (AC 7)	7.25	7	3.58	98.40	24
Malviya Nagar (AC 8)	112.09	15	64.52	200.58	31
Hauz Khas (AC 9)	89.74	8	61.14	70.30	15
Total	45.30	105	50.69	1,178.74	197

Source: Author.

the fund allocation and the actual spending, respectively. Cost of a single work had a wide range from ₹0.25 lakhs to ₹25.0 lakhs; "up to date expenditure" of one project, too, had a similar range between ₹0.15 lakhs and ₹20.23 lakhs. About 90 percent of the projects (N) with 83.6 percent of the total cost of work have been done under the civil department, which indicates the kind of projects undertaken.

About 70 percent of the number of projects (valid $N = 696$) and 71 percent of the total "cost of work" have been taken up in the upper half of the strata (Strata I and II). Interestingly, about 77 percent of the "up to date expenditure," too, have been made in the upper half of the strata (Strata I and II) (Table 8.5). All these show that the allocation of the political funds as well as the actual spending through the MCD has been higher in the upper half of the strata (Figure 8.2). The percentage of implementation, calculated by the ratio of the "up to date expenditure" and the "cost of work," has been clearly higher for the higher strata (Table 8.5). Also, ANOVA results show that the strata has highly significant effect on the overall MLALAD funding [$F(3,688) = 3.287$; $p < 0.05$].[4] Among four ACs, Hauz Khas has utilized the MLALAD fund most in these projects (Table 8.6). ANOVA results also reveal that the ACs have highly significant effect on the overall MLALAD funding [$F(3,688) = 15.721$; $p < 0.00001$].

Table 8.5:
Strata-wise Spending Pattern of the MLALAD Funds through MCD (2004–2005 to 2007–2008)

	Up to Date Expenditure (Valid N=No. of Projects=490)			Cost of Work (Valid N=No. of Projects=696)			
Strata	(P) Total Amount (in lakhs)	Total Amount (%)	No. of Projects (%)	(Q) Total Amount (in lakhs)	Total Amount (%)	No. of Projects (%)	Percentage of Implementation= (P/Q)* 100
I	378.08	34.7	35.3	504.00	27.8	29.5	75.02
II	458.12	42.0	39.8	789.58	43.5	40.3	58.02
III	213.85	19.6	21.4	421.78	23.2	25.1	50.70
IV	40.34	3.7	3.2	100.42	5.5	5.1	40.17
Total	1,090.39	100.0	100.0	1,815.79	100.0	100.0	60.05

Source: Author.
Note: The MLALAD fund spent through MCD has been observed across 78 colonies within four pre-delimited ACs of Hauz Khas, Kalkaji, Malviya Nagar, and Okhla, in which the study area, discussed in the Chapter 7 on water service delivery, is also located.

Figure 8.2:
Strata-wise Pattern of Spending of the MLALAD Fund through MCD (2004–2005 to 2007–2008)

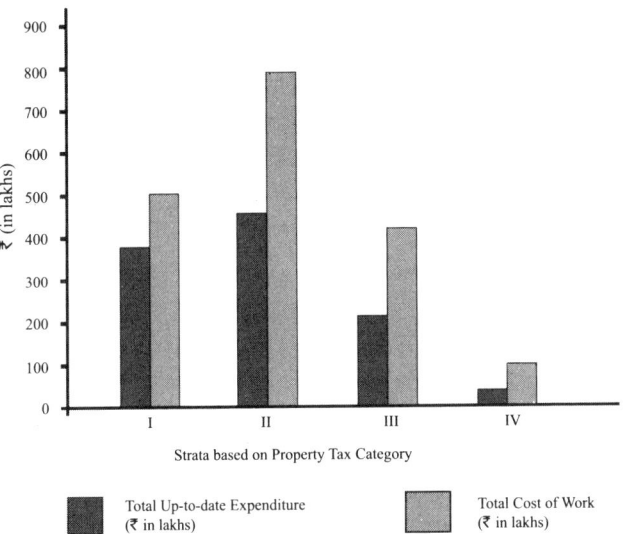

Source: Author.

Note: The MLALAD fund spent through MCD has been observed across 78 colonies within four pre-delimited ACs of Hauz Khas, Kalkaji, Malviya Nagar, and Okhla, in which the study area, discussed in the Chapter 7 on water service delivery, is also located.

Table 8.6:
Assembly-wise Spending Pattern of the MLALAD Funds through MCD (2004–2005 to 2007–2008)

Assembly Constituency	Up to Date Expenditure (Valid N=490)		Cost of Work (Valid N=696)		% of Implementation= (P/Q)*100
	Total Amount (in lakhs)	% of Total Amount	Total Amount (in lakhs)	% of Total Amount	
Hauz Khas	505.66	46.4	587.76	32.4	86.03
Malviya Nagar	276.43	25.4	415.48	22.9	66.53
Kalkaji	204.80	18.8	485.53	26.7	42.18
Okhla	103.49	9.5	327.03	18.0	31.65
Total	1,090.39	100	1,815.79	100	60.05

Source: Author.

Note: The MLALAD fund spent through MCD has been observed across 78 colonies within four pre-delimited ACs of Hauz Khas, Kalkaji, Malviya Nagar, and Okhla, in which the study area, discussed in the Chapter 7 on water service delivery, is also located.

Time of Spending of the MLALAD Fund

During the study period, elections were held at all three levels, Parliament election in 2004 and 2009, Delhi Assembly election in 2008, and the MCD election in 2007. With more than 50 percent of the funds allocated and spent during election years, one may see the selectivity in the timing of the spending of funds (Figures 8.3 and 8.4).

Figure 8.3:
Year-wise Spending of the MLALAD Fund through DJB (2004–2009)

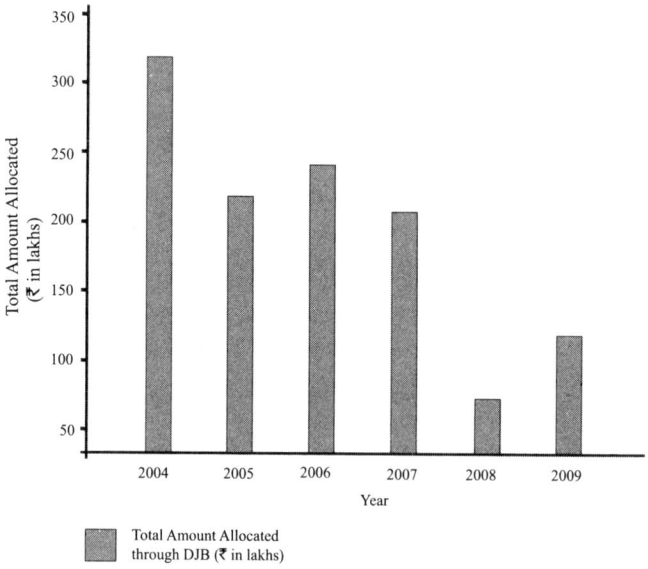

Source: Author.

Through the DJB, there has been a tendency to allocate more funds around election years. About 51 percent of the total amount was allocated and 54 percent of the total number of projects was introduced during election years. During the study period (2004–2009), the MLALAD spending was steadily less as the tenure goes by and was the least at the Assembly election year of 2008. Despite the Election Commission's guidelines on the code of conduct, there has been ample time for the utilization of the fund since the

Figure 8.4:
Year-wise Spending of the MLALAD Fund through MCD (2004–2005 to 2007–2008)

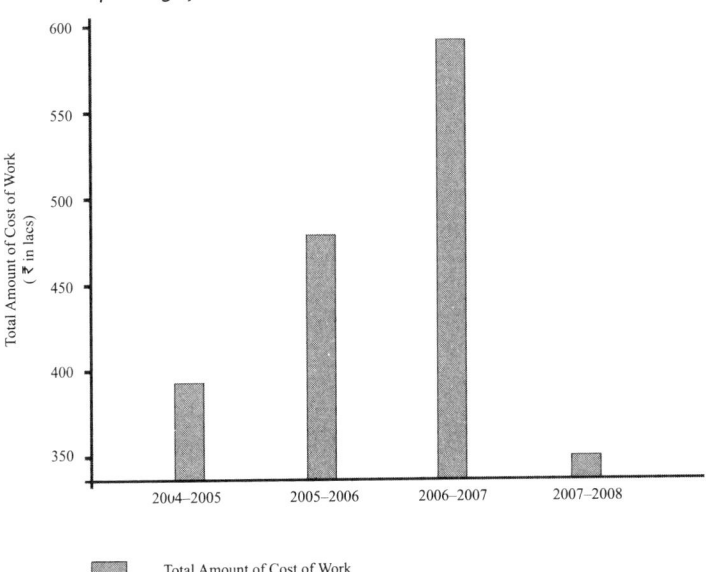

Total Amount of Cost of Work
through MCD (₹ in lakhs)

Source: Author.

Note: The MLALAD fund spent through MCD has been observed across 78 colonies within four pre-delimited ACs of Hauz Khas, Kalkaji, Malviya Nagar, and Okhla, in which the study area, discussed in the Chapter 7 on water service delivery, is also located.

election took place toward the end of the year. Through the MCD, about 53 percent of the total number of projects was introduced in all the election years together and about 54 percent of money was spent in those years too. Maximum amount was allocated in the year 2006–2007, around the time when the MCD election took place.

Remarks

Established notions, like the *vote bank* related spending or, the *patron–client* relationship, that politicians are supposed to enjoy with the lower socioeconomic strata and spaces of society, both, seem limited in explaining the patterns of the MLALAD funding here. Another apprehension is: Besides fetching more votes (*vote bank*), one can pass on low-quality works in these areas. However, qualitative assessment of implemented projects to support such nervousness needs separate analysis and is worth taking up. Also, the assumption that the politics of distribution is coerced by the *coalition* between the politicians and the *influential* people and, in turn, favors the rich is not so conclusive. Agency-wise spending pattern has certain differences. More funds were allocated through DJB, to the lower half of the socioeconomic spaces, whereas through the MCD, it was to the upper half of the society. All these observations may well be perceived as manifestations of the multiple strategies and politics of governmentality.

*   *   *   *   *   *

Politics of Distribution in Water Supply: A Summary of Empirical Findings

Socioeconomic Spaces and the Property Rights

Question: What is the correspondence between legal property rights and socioeconomic spaces?

1. Empirical findings:

   a. Property tax strata have significant effect on the change in income, expenditure, and savings as well as on the combined "socioeconomic factor."

   b. Strata-wise mean value of the "socioeconomic factor," based on income, expenditure, and savings, clearly indicates that the higher strata have higher socioeconomic factor.

   c. Average monthly income is higher in higher property tax strata. Average monthly income of Strata I (₹86,405.84) is about seven times that of Strata IV (₹12,432.43).

   d. Average monthly expenditure on important heads, such as education, electricity, transport, and entertainment, which are about 25 percent of the overall average income, indicates higher expenditure in higher strata.

e. Average monthly savings are higher in higher strata too. In fact, the average monthly savings in Strata I is almost nine times that of Strata IV.

2. Interpretations:

a. Higher property tax strata represent spaces occupied by higher socioeconomic strata.

b. In the absence of any direct measure of socioeconomic spaces, property tax strata, based on the tax categories set by the MCD, can be used as the appropriate surrogate indicator to classify the socioeconomic spaces of the residential setting of Delhi.

Water Supply and Socioeconomic Spaces

Question: How does the delivery of urban water supply in Delhi correlate with city's socioeconomic spaces and its legal property rights?

1. Empirical findings

a. There are gaps between the intention and the actual delivery of water supply.

- Different claims on the coverage of DJB supply, ranging from 84 percent to 100 percent, are made by different government agencies.
- Disparity exists between the planning intentions and its implementation in water supply.
- Demand–supply projections do not conform to the claims of coverage by agencies, such as, DDA or DJB.
- Demand–supply gap gets even wider due to NRW. Different government and private agencies provide different figures, varying from 24 percent to 45 per cent, for the "loss of water" in the distribution process. Such variable estimates of the NRW provided by different interest groups somewhat underlines the redundancy of the projected NRW targets by the DJB.

b. Politics of information, formed out of ambiguities surrounding the factual difference on planned distribution and actual delivery of water supply, makes the claim of cent percent coverage hard to believe.

c. Horizontal Inequity exists across geographical spaces in the city, since the NCRPB (1999) presented a map with territories having different quantity of supply.

d. Inequity prevails due to differential policy norms.

- Per capita policy norms and benchmarks differ across government agencies, such as the DDA, the CPHEEO, and the DJB, ranging from 135 lpcd to 225 lpcd. Differential policy norms for water supply are based on the "land tenure"/settlement types. DJB's responsibility to provide individual water supply to around 20 percent of the population, living in JJ clusters and unauthorized colonies, seems absent leaving all these people dependent on hand-pumps or community taps installed by the DJB.

- Politics of information due to varying estimates of the number of informal settlements tends to affect the population projections, the city level average requirement for per capita calculations, overall water demand, and corresponding supply.

Out of necessity to supplement the lack of the state provision of water supply, slum-dwellers are often found paying a "poverty premium" of four to seven times more than the DJB price (Dimri and Sharma 2006).

e. *Vertical inequity* occurs in the delivery across socioeconomic spaces of legal colonies.

- Socioeconomic strata have significant effect on water supply. As a result, the delivery of water varies significantly across the socioeconomic strata.

- The higher the tax strata, that is, the socioeconomic strata, the higher the quantity of water supply. Two distinct slabs of water supply emerge, one, consisting the top two strata (I and II) the other, the bottom two (III and IV). The topmost strata seems the most privileged one in terms of the delivery of water supply and enjoys the filtered water supply of around 13 percent more than the recommended standard of 135 lpcd.

- Users from both the extreme ends of the strata (I and IV) seem more satisfied with the delivery of water supply including its quality, whereas users from the mid-level strata (II and III) have lower LoS with the supply. One of the reasons for the top strata being satisfied with the quantity of water supply may be the supplementary sources of water other than the piped supply available to them. Despite lower DJB supply, lower socioeconomic groups have higher LoS as they seem to be used to the inadequate service delivery and, also, have very little expectation from the agencies to provide better quality of water as well.
- Lower LoS of the people from the mid-strata of II and III somewhat underlines a generic phenomenon of social dissatisfaction of the MIG with the basic service delivery.

f. Relationshi*ps between water supply, territories of governance, and strata*

- Aspects of water supply are affected significantly across the strata even when seen in relation with territories of governance, and higher strata are more privileged in terms of the delivery of water supply.
- Delivery of water supply changes significantly across ACs.
- As seen in the surveyed area, DJB service zones almost have no effect on the variation of water supply.

2. Interpretations:

a. Same supply norms and legal property rights do not necessarily ensure equal water supply, and the real delivery varies across socioeconomic spaces within planned residential areas in a way that the higher strata of socioeconomic spaces (i.e., the colonies) get higher supply of water.
b. Inequity across socioeconomic groups clearly shows that in reality, water supply across the colonies (including the planned ones) with legal property rights has multiple delivery situations.
c. It appears that supplying a scarce resource, like water, more to the richer sections of the society even amongst the so-called legal colonies is not a policy decision, but a governmentality technique.

d. Such a technique reveals the strategic selectivity of the state in favor of the higher strata of the society and the spaces occupied by them in the city.

Water Supply, Political Funding, and Socioeconomic Spaces

Question: How does political patronage relate to the socioeconomic spaces (indicated by the property tax strata) in Delhi?

1. Empirical findings:

a. Spending of the MLALAD funds through DJB

- ACs have significant effect on both the total funding and the per capita funding of the MLALAD through DJB.
- Socioeconomic strata have significant effect on the spending of the MLALAD funds through the DJB. Consequently, overall spending of the political funds through DJB varies significantly across strata.
- Interestingly, political funding for water is more for the colonies where the lower half of the strata live (Strata III and IV), but in effect, per capita fund utilization is relatively more in the higher strata.
- The allocation, in other words, the redistribution of the fund is higher among the lower half of the socioeconomic strata and lower among the upper half, 68.33 percent and 31.67 percent, respectively.
- Per capita spending of the fund may actually benefit the higher strata more when average per capita (MLALAD) allocation is higher in spaces occupied by the higher strata of society. Per capita allocation of the MLALAD fund, too, has a significant positive correlation with the total allocation of funds ($p < 0.01$).
- There is a tendency to allocate smaller amount per project on water supply and cover more number of colonies. The amount allocated for around 75 percent of number of projects has been less than ₹8.50 lakhs and average project cost has been about ₹6.0 lakhs.

b. Spending of MLALAD funds through the MCD

- Although MCD cannot spend the MLALAD funds on water supply, yet about 90 percent of the total MLALAD fund is used through the agency. Hence, to observe patterns of the spending of the political funds in general, the utilization of the MLALAD funds through the MCD is analyzed.
- ACs have highly significant effect on the overall MLALAD funding through the MCD.
- Socioeconomic strata have significant effect on the overall MLALAD funding. Hence, the funding varies significantly across strata.
- The allocation of funds and the actual spending through the MCD is higher in the upper half of the strata. About 70 percent of the number of projects, the cost of which amounting to about 70 percent of the total cost of work, have been directed to the upper half of the strata.
- In terms of actual spending, about 77 percent of the "up to date expenditure," too, have been made in the upper half of the strata (Strata I and II).
- The percentage of implementation, calculated by the ratio of the "up to date expenditure" and the "cost of work," is clearly higher for the higher strata. Strata I has the highest percentage of implementation (about 75 percent) and Strata IV, the lowest (about 40 percent).
- The timing of the utilization of the funds, as expected, seems to be chosen selectively, keeping in mind the election years: More than 50 percent of the funds are allocated and spent during the election years.

2. Interpretations:

a. Multiple techniques of governmentality influence practices of political patronage reflected in the spending of political funds that overwhelmingly favor neither the higher nor the lower socioeconomic strata, yet have tendencies to favor spaces occupied by richer sections of society.

b. Spending more on the projects in the lower half of the socioeconomic spaces to benefit the larger section of people is possibly expected in a patron–client relationship that politicians may have had with their electorate.

c. Such techniques of governmentality are typical of the politicians who try to appease larger sections of the population and aim to project themselves as patrons.

Notes

1. Under the MLA Fund, MLAs allocate funds to DJB. DJB in its official capacity decides on the projects within the concerned MLA's AC where the amount would be spent and MLAs can always give their suggestions on the projects. Under the MLA Priority Fund, improvement works recommended by each MLA are taken up by the DJB.

2. Since the dependent variable, amount allocated (in lakhs) of MLALAD, does not have a normal distribution, a logarithmic transformation is made. The transformed variable has a normal distribution and can be subjected to ANOVA.

3. The mean value of the funding is the highest at the lowest end of strata. Strata IV has a significantly higher mean value than the rest of the strata ($p < 0.05$). Mean value of funding is not important here because it only gives total amount/no. of projects, that is, the average value of projects and, hence, does not necessarily indicate the funding pattern vis-à-vis the strata.

4. Since the dependent variable, "cost of work" of the MLALAD, does not have normal distribution, it has been transformed to its logarithmic values. The transformed variable has normal distribution and, therefore, ANOVA can be conducted with it. ANOVA results show that the strata has highly significant effect on the transformed dependent variable [$F(3,688) = 3.287$; $p < 0.05$] and, therefore, on overall MLALAD funding.

PART III

Assorted City

9

Politics of Practices, Its Ambiguity, and Governmentality

So far, two broad perspectives are opened up: one, on the notion of equity and justice in the distribution and delivery of basic urban services, such as water supply in Delhi and, the other, on possibilities of city reading through the lens of justice and equity. Empirical part of the research brings out certain key points to understand existing conditions of justice and equity in the provision of water supply in Delhi. Inequitable service delivery is created when

- disparate access across socioeconomic spaces in the city occurs out of differential policy norms,
- despite same norms, actual delivery differs leading to conditions of *horizontal inequity* across geographical spaces in the city, and
- despite same norms or access to the *network* or *means*, actual delivery differs creating situations of *vertical inequity* across socioeconomic spaces of legal colonies (Figure 9.1).

Certain nuances of these inequities in the service delivery may now be interpreted in light of the theoretical notions discussed earlier. Conditions of *horizontal inequity* can be seen in light of "location-specific distributive decisions," where accessibility to and availability of resources and services vary because of the location of the users from the main source of distribution and its network. Inequity in service delivery, created by the disparate access to resources and services across socioeconomic spaces out of "differential policy norms," based on the settlement type or the legal status of the colonies, underlines that instead of the locational proximity to the supply network of such services, the provision of *means* to the user and the user's accessibility to the network

Figure 9.1:
Equity and Justice: Observations and Possible Approaches

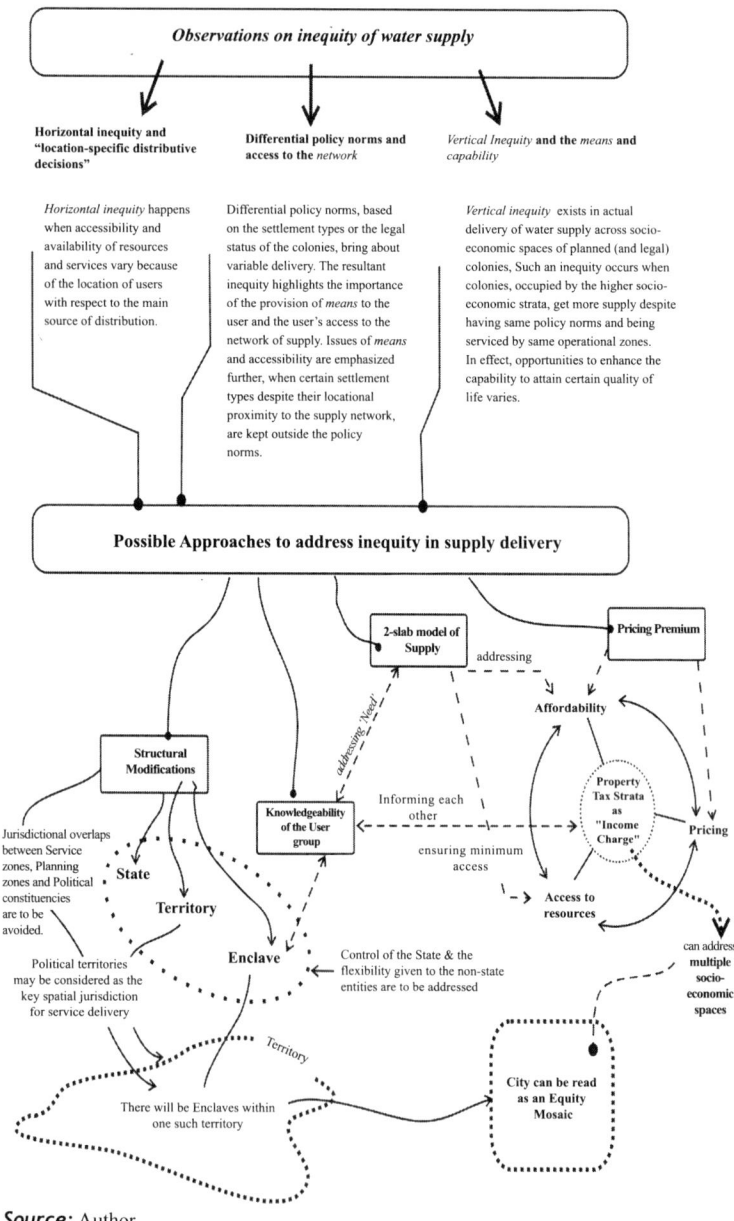

Observations on inequity of water supply

Horizontal inequity and "location-specific distributive decisions"

Differential policy norms and access to the *network*

Vertical Inequity **and the** *means* **and** *capability*

Horizontal inequity happens when accessibility and availability of resources and services vary because of the location of users with respect to the main source of distribution.

Differential policy norms, based on the settlement types or the legal status of the colonies, bring about variable delivery. The resultant inequity highlights the importance of the provision of *means* to the user and the user's access to the network of supply. Issues of *means* and accessibility are emphasized further, when certain settlement types despite their locational proximity to the supply network, are kept outside the policy norms.

Vertical inequity exists in actual delivery of water supply across socio-economic spaces of planned (and legal) colonies, Such an inequity occurs when colonies, occupied by the higher socio-economic strata, get more supply despite having same policy norms and being serviced by same operational zones. In effect, opportunities to enhance the capability to attain certain quality of life varies.

Possible Approaches to address inequity in supply delivery

2-slab model of Supply

addressing

Pricing Premium

Structural Modifications

addressing 'Need'

Affordability

Knowledgeability of the User group

Informing each other

Property Tax Strata as "Income Charge"

Pricing

Jurisdictional overlaps between Service zones, Planning zones and Political constituencies are to be avoided.

State

Territory

ensuring minimum access

Access to resources

Political territories may be considered as the key spatial jurisdiction for service delivery

Enclave

Control of the State & the flexibility given to the non-state entities are to be addressed

can address multiple socio-economic spaces

Territory

There will be Enclaves within one such territory

City can be read as an Equity Mosaic

Source: Author.

of supply are of importance. *Vertical inequity* across socioeconomic spaces of legal colonies reveals that sharing the same provision of means or accessing the same network, too, does not ensure the same delivery and as a result, opportunities to enhance the capability of attaining certain quality of life vary.

These observations contribute in developing possible approaches to deal with certain issues of inequity in supply, and in that direction, I propose the following fourfold suggestions (Figure 9.1):

- Policy norms of water supply may have the *two-slab model* which includes *the minimum access to supply* equally across all sections of society and *the differential optimal access to supply* varying across the strata depending on the need, consumption pattern, and affordability of different socioeconomic strata.
- *Pricing premiums* are to be charged in a way that the richer people pay more for services.
- *Knowledgeability of the user group* is to be utilized as the *bottom-up* information system to calculate the need, which will address issues of accessibility–affordability–pricing and, therefore, what Harvey (1975[1973]) terms as, the "income charge."
- *Structural modifications* are to be envisaged in order to explore options of control and flexibility between governing entities (e.g., the state and/or non-state actors) in spatial and operational terms.

Also, highlighted are the politics of practices, its ambiguity, and governmentality, which create an *equity mosaic* depicting multiple conditions of living in the city. While reading the city, I recognize such multiple shades of existence.

Politics of Distribution in Urban Water Supply: Multiple Shades and Conditions

Observations on empirical findings tend to reveal differences that the real delivery has with the policy intention and the larger ideology of just distribution. However, inequities existing in the real delivery of water supply implicate each other in indirect ways.

Horizontal Inequity and "Location-specific Distributive Decisions"

The first level of politics of distribution happens due to conditions of *horizontal inequity*, when the accessibility and the availability of resources and services vary because of the distance of the location of the users from the main source of distribution. As a result, uneven distributional territories are created. In Delhi, such situations exist at the larger city level. A common sense understanding would blame the lack of technology for such gaps in distribution. At the same time, the governmentality of the state in persisting with such territorial deficits may not be overlooked. Broader Marxist critiques, of which arguments of Neil Smith and David Harvey are particularly significant, have underlined such phenomenon in great details.

Differential Policy Norms and Access to the "Network"

Differential policy norms, based on the settlement types or the legal status of the colonies, bring about variable delivery in reality. The resultant inequity in the service delivery is formed out of disparate accesses to resources and services across socioeconomic spaces. It highlights the importance of the provision of the *means* to the user and the user's accessibility to the network of supply, and not the locational proximity to the supply network of such services. Within a service zone for water supply, the proximity and distance should matter little to the provision of water supply. Whether one is in the network is perhaps more important than whether one is close to the network. For example, one near water supply lines and networks may not necessarily enjoy the supply, whereas someone, having the connection of the network, can access the supply (Figure 9.2).[1]

Access to the piped water supply by the urban poor may also be an example to explain the access–network relationships. Many squatter settlements, despite being located near the physical networks of water supply or situated next to the colonies served by such networks of water supply, do not have any access to the piped water supply. Besides that, the state, in this case the CPHEEO (1999) under the MoUD of the Government of India, recommends differential policy norms for water supply across people and spaces based on certain assumptions

Figure 9.2:
Access and Network

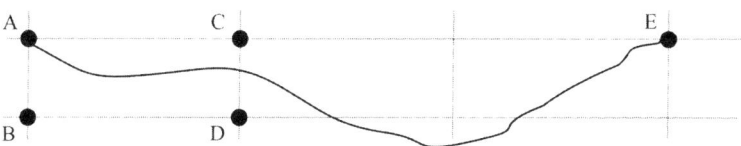

Source: Adapted from Latour (1996).

Note: Latour (1996) explains that "A is further from B than from E once connections are taken into account; the proximity of A and B or of C and D is due to the connections established by the grid system and the work of geographers..."

on the user pattern and the settlement types. Accordingly, settlements existing outside the legalities of the planning (or policy) documents are considered in the *state of exception* and are, by and large, subject to the less water supply to begin with. Governmentality at this instance is not necessarily dictated by the locational notions of proximity (and distance) that urban geographers, like Harvey (1975[1973]), mention.[2]

Recommendations of the norms are the policy framework based on certain agreements on the quantity and the pattern of anticipated usage of water—this is the basis of social contract in Rawls' theory of justice. In Delhi, we see a disagreement between norms set by two agencies. On one side, the CPHEEO and the MoUD of the Government of India under which the DDA operates, and on the other side, the DJB under the Government of National Territory of Delhi. The DJB suggests a uniform supply across all settlements of the society, whereas the CPHEEO/MoUD recommendations indicate a differential delivery based on the legal status of the colony.

Policy norms are essentially used as the benchmark for calculating water demands. But among all the four strata of socioeconomic spaces, one could see in the empirical research that only the top strata of the society have actually received more water than the quantity recommended by the norms. Also, the "Willingness to Pay" (WtP) of the PWC study (PWC et al. 2004) indicates that the preferred quantity of water in the slums (81 lpcd) is more than the prescribed norms of the CPHEEO (70 lpcd), whereas that in the planned colonies (155 lpcd) is less than the norms (225 lpcd).

Also, there exist politics of information on a number of aspects including the coverage of water supply, the loss of water in distribution, and the number of unauthorized and squatter settlements. All these result in inaccurate descriptions and estimates of the supply in reality, and consequently, the demand–supply calculation becomes ambiguous and unreliable.

However, at the end, the policy norms and the availability of the network of supply, one can see in the empirical research, do not guarantee the actual delivery, neither in the legal colonies nor in the settlements existing outside, what Holston (1998: 253) says, the "text-based rights" ascribed by the planning documents. In fact, the accepted policy norms are supposed to encourage the just or the equitable distribution, especially across the "least advantaged" people of the society or spaces (Harvey 1975[1973], 2003[1992], Rawls 1969, 1971, 1993). But, when these groups of people and their spaces are deprived of basic services like water supply, the agreement on justice and equity is compromised. As a result, poorer people, generally less capable of compensating for these shortfalls, suffer the most.

The very concept behind the projected usage of water, therefore, is to be revisited. The present norms recommend quantities, which are difficult to match in reality. When limited availability of water is to be rationed, norms must recommend more realistic quantity of supply based on the actual delivery situation for the demand calculation. The present reality of the shortage of water for distribution manipulates the technique of governmentality resulting in higher water supply to higher strata.

In a generic sense, realities of scarcity lead to the governmentality of inequitable distributions. Such realities, too, contribute to the change in the worldview, which in turn, influences the ideology. The ideology, then, modifies its original position of the democratic welfare state. This is a reiteration of the conceptual relationship in the formulation of the policy (Figure 9.3).

"Vertical Inequity" and the "Means" and "Capability"

Vertical inequity across socioeconomic spaces of legal colonies reveals that sharing the same provision of means or accessing the same network may not ensure the same delivery in reality. In effect, opportunities to enhance the capability of attaining certain quality of life would vary. This poses a critical twist to the explanation of conditions of inequity in terms of access and network. The curious question is: What does it signify when the delivery of water supply varies across the socioeconomic strata

Figure 9.3:
IPG Relationships with Reference to Urban Water Supply in Delhi

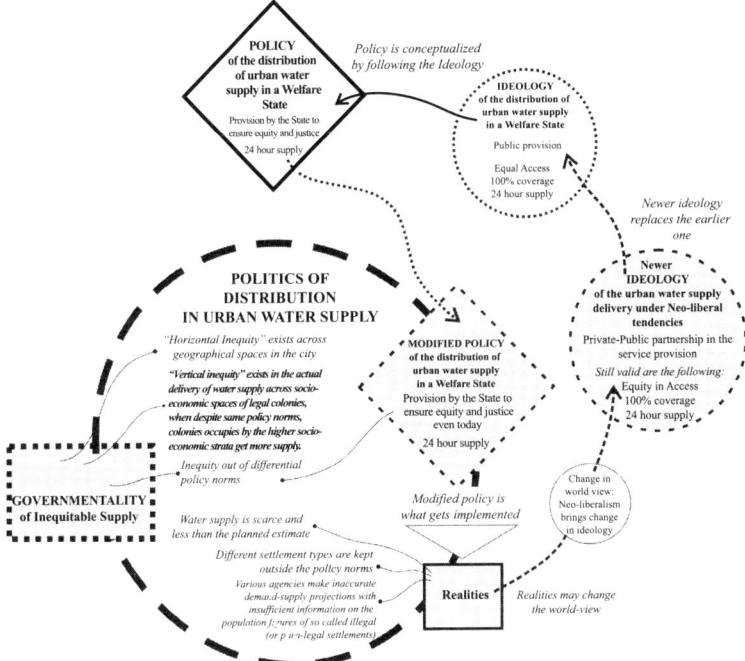

Source: Author.

Note: This diagram can be seen with reference to Figure 1.1, which establishes a generic framework for the overall relationship between IPG. Here, the framework is made topical and contextual by being specific to water supply in Delhi. Politics of distribution happens, I perceive, out of inequitable provision at various scale-levels, further implicated by ambiguous information on actual situations.

of the planned colonies situated within the *same* water supply OZ and serviced by the *same* UGR with the *same* policy norms?

In such cases, inequity in the supply situation can be at two levels, one, at the *means* of supply and the other, at the *capability*. Planned colonies in the MPD are somewhat conceived as the settlement types with complete facilities of life and as a result, a policy of higher per capita water supply norms is recommended for such colonies. Having taken clues from a certain established line of thinking, I have associated the planned situation with the legal state of condition and, therefore, with the notion of citizenship (Bhan 2009, Chatterjee 2004, Holston 1998, 2008, Holston and Appadurai 1996, Marshall 1950). The assumption is, citizenship is a status to be enjoyed by the "full members of a community" and all citizens "are equal with respect to the rights and duties with which the status is endowed" (Holston 1998, Marshall 1950: 92). Hence, people across all the planned colonies are supposed to enjoy such a status guided by the planning policy. Instead, the empirical work points out two clear-cut slabs of supply, within which the higher strata enjoy more water.

A possible explanation of the ground reality of such differential deliveries would be that the *means*, to use the Rawlsian phrase, do not necessitate the actual delivery and the consumption at the end. This is one of the major points of departure, I have discussed earlier, from John Rawls' theory of justice (1971) to what Amartya Sen elucidates in *The Idea of Justice* (2009). Means, here, stand for the norms and policies, networks and availability of resources in the network, etc. For example, a stand-pipe for the supply of drinking water is provided in a colony, but if the quality of water from that tap is not fit for drinking, or the water is less in the tap, the provision of the means of supply would not meet the objective.

Capability, actual opportunities, and "substantive freedom" are essential to access the means and fulfill the ends (Sen 2009). However, as the fundamental precondition to all that, societies must provide a person with the means, that is, the opportunities of facilities and services, which in this case is water supply, and without the means, he does not have the capability to enhance his conditions and, hence, shall be considered deprived (Rawls 1971, Sen 2009).

All versions of inequity in water supply, as observed empirically, clearly indicate differences in the opportunities offered. What is further disturbing is the lesser prospect for the lower strata of society, as seen in the planned colonies in the study area, in accessing the means supplied by the state!

Now the question is how does one work toward realistic considerations for any possibilities of an equitable supply?

Possible Approaches to Address Inequity in Water Supply

I do not intend to propose any grand *vision* here to achieve a *just city* although one acknowledges Fainstein's (2005: 127) argument that the *just city* is to be considered as the "object of planning theory." The intertwined relationship of the critique and the vision toward overcoming the injustice in the city is certainly worth a mention here:

> Utopian thinking has two moments that are inextricably joined: critique and constructive vision. The critique is of certain aspects of our present condition: injustice, oppression, ecological devastation to name just a few. . . . Moral outrage over an injustice suggests that we have a sense of justice, inarticulate though it may be. . . . If injustice is to be corrected . . . we will need the concrete imagery of utopian thinking to propose steps that would bring us a little closer to a more just world. (Friedmann 2002: 104; quoted in Fainstein 2005: 127)

Here, it is felt appropriate to leave such aspirations and debates to more experienced scholars (Fainstein 2005, 2005[2004], 2006, Healey 2003, Sandercock 1997). However, the critical commentary I have posed so far leads to certain corrective measures. With such intentions, concepts are suggested to deal with issues involving *access to supply*, *pricing*, tapping the *knowledge* of the users as the *bottom-up* information system to frame the policy and the necessary *structural modifications* to accommodate all these changes.

Two-slab Model of Supply: The Minimum Access and the Optimal Access to Supply

A possible approach can be the adoption of a *Minimum Access Model* to ensure minimum supply equally across all socioeconomic groups. Such a model is based on two categories in terms of quantity of supply: minimum supply and optimum supply. Once minimum supply is made certain for all socioeconomic groups, the delivery of optimum supply can be worked out in two broad ways depending on the ideology of the state. For a welfare state, the per capita optimum supply should be equal across all sections of the society, whereas, for a neoliberal state, per capita optimum supply may follow a differential scale depending on the need, consumption pattern, and affordability based on the socioeconomic strata of a colony and eventually, may become higher for the higher strata. Looking at the present ideological direction of the Indian state, one may be open to the policy option of the differential scale of optimum supply. However, such difference in distribution would be a stated policy and not an outcome of unsolicited governmentality as it exists today.

This two-slab supply conceptually differs from the existing delivery norms of the CPHEEO (and MoUD) and the pricing slabs of the DJB. Prevalent delivery norms of the supply of the CPHEEO and MoUD recommend the variable supply ranging from 70 lpcd for squatter settlements and 225 lpcd for planned colonies, but the two-slab supply of the minimum access model must guarantee the complete coverage of the minimum supply across all the strata after which the optimum supply is to be provided.

Since the purpose of this work is to neither quantify the equitable supply nor ascertain the just pricing of water, no attempt is made here to determine the quantity of minimum supply or the exact pricing of water. Instead, I will only explore a conceptual explanation with the help of hypothetical notations:

> If the minimum supply quantity=x; no. of people to whom supply is to be made=n; total quantity for minimum supply=nx. To begin with, nx will have to be distributed equally (x) across people in all strata. If the total quantity of water available=y, quantity of water available for optimum supply=y–nx, which will be made available for optimum supply for differential distribution. The value of y will vary across the year or even on a daily basis. Based on the survey on the quantity and pattern of water consumption across all strata, one may possibly approximate the strata-wise requirements of the supply.

In contrast to the variable delivery norms across settlement types identified by the planning policy, this two-slab notion appears to suggest a delivery pattern across all socioeconomic strata by ensuring equal minimum access and opportunity for all. The daunting question is: How will the optimum supply be determined? Will it be demand-based? If so, how will the required information be gathered? I will try to address some of these questions.

Pricing Premium: Making Richer People Pay More for Services

Differential deliveries, one could see earlier, exist as a coercive technique of governmentality, essentially arising out of the state's inability to achieve unrealistic high norms of supply across all sections of society within the limited supply condition. Consequently, the state of exception is formed out of the disparity in water supply which, more often than not, is higher for higher strata. Also, people living in slums and squatter settlements mostly have very limited access (in the form of tanker supply or the stand-pipe supply) to the institutionalized supply system. Sometimes, they have no access as well. To reduce the deficit in supply, such section of people, quite ironically, are compelled to pay even more; for example, a study by Dimri and Sharma (2006), explained elsewhere in this work, show that people living in the notified slum areas in South Delhi tend to pay on the price of water a "poverty premium" of four to seven times more than that of the DJB. Since poorer people, in general, are less capable to mitigate the shortfalls, such delivery pattern appears unjust.[3]

Instead, the pricing of supply may be worked out by charging a premium which makes people from the higher strata pay more for per unit of supplied water. This pricing strategy for optimum may incorporate the pricing system of the DJB, based on certain slabs of supply, which require revision as per the quantity of the minimum supply.[4]

The settlements outside the property tax coverage, for example, squatter settlements or unauthorized colonies may be considered as the base level and all other four strata (that includes eight tax zones as elaborated earlier) can have differential unit price of water. Colonies outside the property tax strata, arguably, should not be charged a premium, but the rest should pay *a premium variable with the strata.*

The amount of the premium is, indeed, a separate exercise, and may be worked out on the basis of sample surveys across all socioeconomic strata. Certain factors, such as, the consumption quantity and billing, willingness to pay and income-expenditure patterns of different socioeconomic strata, total house tax paid, etc., can be taken as indicators of the premium.

At the time of finishing the final manuscript of this book, Delhi got a new government of Aam Aadmi Party (AAP), a recent political entity headed by Arvind Kejriwal. Albeit the government stayed in the office for a short while, it introduced a revised tariff system worth a mention here. According to this new tariff, 20 kiloliters of water per month per family (or an average of about 132 lpcd for a five member family) will be supplied free of cost to areas apart from NDMC areas, Delhi Cantonment, and Dwarka. If the consumption goes above 20 kiloliters, the entire water consumed will be billed. At the same time, the water rate would be increased by 10 percent. Newspaper reports observe that such a decision may underline "the rich versus poor debate" by benefitting nine lakh families, whereas for the remaining six lakh families water would become more expensive (Lalchandani 2013). Since the government has resigned, this tariff structure is also withdrawn, at least for the time being. This tariff structure, indeed, stresses upon a premium on consumption of water.

To come back to the main discussion, both the pricing and the two-slab model of the access to supply require certain clarifications on how the information on the demand of water may be gathered.

Water Demand Calculations: Potential of the Knowledgeability of the User Groups

A critical problem would be to calculate the total water demand based on the per capita requirement for a particular colony, the socioeconomic space, accuracy of which depends on that of the population of the colony. At the city level, the demand calculation seems difficult with ambiguity surrounding population figures due to considerable variations in the number of slums and squatters calculated by different agencies. Is there any alternative way to work out the quantity delivered? A couple of thoughts are shared here.

First, is it possible for the resident bodies (like the RWAs) to give a monthly requirement for the colony or the settlement? Such a bottom-up and decentralized approach to get demand requirements may lead to more realistic estimates.[5] A parallel comparison would be the self-assessment of the house tax in Delhi, which has made property taxation easier. Similarly, the advocacy, awareness programs, and a simple system of demand calculations for water supply may also be attempted.

Second, for the self-assessment or the community based approach of the demand calculations, should the total covered area equivalent to that of the house tax be taken as a factor of the water demand? A common sense observation suggests that the size of a house influences the total quantity of water required for the household, because for a larger house, more water tends to get used for floor washing and for flushing the toilets. MPD 2021 mentions that about 40 percent of the water tends to get used for these two purposes (DDA 2007a: 148). Hence, an average of total covered area per house may be worked out as a representative figure for a colony based on the sample survey or from the available property tax data.

The probable question is: How does one accommodate these notions within an operational mechanism, while working toward an equitable supply?

Structural Modifications: Possibilities of "Welfare Provisions"

This work does not claim to give any ready *ideal* solution to eradicate current conditions of inequity. Instead, it attempts to bring out certain dimensions of equitable distribution and related nuances pertaining to water supply. During summer months in Delhi, remarks like the following one by the then Chief Minister of Delhi, Mrs Sheila Dikshit, are quite common in the newspapers:

> [T]his summer there is no additional source of water for Delhi and with population growing, the only option we have before us is better water management practices. *Better pricing and equitable distribution is the only way forward.* (Lalchandani 2011; emphasis included)

So far, broad structural arrangements and strategies to deal with shortcomings of urban water supply in Delhi have been attempted and understood by the state in three ways:

1. Discussions on a proposed regulatory framework having a joint venture between the Delhi government and various private companies (Lalchandani 2011).
2. The "World Bank model" to manage distribution and continuous supply while outsourcing water supply and sewerage services through management contracts and, in turn, effecting an increase in the tariff to recover 80 percent of the operation and maintenance costs (Bhaduri and Kejriwal 2005, DJB 2004, Maria 2008, Parivartan ca. 2005).
3. The "state hydraulic model," mentioned earlier, laying down the planning norms and policy as well as augmenting the water sources to address the demand–supply gap (Bakker 2005, Barraqué 2004, Maria 2008).

The policy of water supply in Delhi has followed the "state hydraulic model" till now. The estimates as per the MPD and the DJB have calculated the water demand by multiplying supply norms with population projections. In order to cope with the rise in population in Delhi, policy recommendations have often relied upon the long-term goals, time-consuming ones for implementation. To increase the capacity of the water treatment plants or to find new sources of water for the city by buildings dams would be examples of such policy moves (Maria 2008). As time passes by and population projections fall short of the actual figure, corresponding projected water demand calculations become less than the real needs and the whole demand–supply planning becomes redundant. These planning policies and practices result in the "supply-led solutions" to satisfy administratively defined needs and goals—a dominant model followed in most of the OECD countries in the 20th century (Bakker 2005, Barraqué 2004, Maria 2008).[6]

More recently, by taking clues from the "World Bank model" for utility management, the DJB prepared a plan for "reforming its water supply distributions" with financial support and technical assistance of the World Bank and the vision was to ensure "universal continuous (24×7) safe water supply and sewerage services in an equitable, efficient, and sustainable manner by a customer oriented and accountable service provider" (Bhaduri and Kejriwal 2005, DJB 2004, Singh and Shukla 2005). The DJB's "Reform Plan" in 2004 outlined two broad strategies: First, an increase in tariff to recover 80 percent of its operation and maintenance costs and second, the outsourcing of water supply and sewerage services through management contracts for the OZ II and III as the Pilot project (The World Bank 2005).[7]

This strategy, however, came under severe criticisms (Bhaduri and Kejriwal 2005).[8] Major reservations were against the increase in the water tariff due to very stiff management fees to meet salaries of employees sent by the multinational companies, high operational expenses to run zones on day-to-day basis, and the unaccountability surrounding the capital investments (Parivartan ca. 2005).[9] Such criticisms resulted in the postponement of the reform agenda of the DJB and the withdrawal of the application for the World Bank loan at the end of the year 2005.

Apparently, the "Regulatory model" was being discussed by the government toward the end of 2011, which intended to formulate a private–public joint venture for "the treatment and distribution of water" (Lalchandani 2011). However, clear implementation of the model has not been reported as yet. It was anticipated in the model that the government agency would regulate the pricing of water and the overall services of the private operators. A newspaper report suggests that such model would require changes in the prevailing water act in Delhi (The Delhi Water Board Act 1998) to evolve a "three-tier system" with the state-controlled Water Board at the top looking after the procurement of raw water, handling of inter-state disputes, and augmentation and management of infrastructure (Lalchandani 2011).[10] Interestingly, the state should continue to own the assets of the Water Board, yet new structural changes would handle equity, efficiency, and accountability. The apparent paradox is that, on the one hand, government bodies in India, in general, are not particularly associated with efficient operations and, on the other hand, the state has to rely upon the accountability of the private agencies to ensure equity for its people. This model, however, seemed to have considered certain issues of horizontal equity, which was to be achieved through the transfer of water between two command areas covered by two separate water treatment plants. In other words, if one water treatment plant had more capacity and the corresponding command zone had adequate supply, the surplus should be routed to other areas having less capacity. In turn, horizontal (geographical) inequities were addressed at the city level. But, such schemes would reportedly need necessary technological adjustments of the present system (Lalchandani 2011). This model as one understands would require more refinement. Also, methods to address the complexity of equity in distribution and in pricing across the socioeconomic groups and spaces have not been made clear. Shortcomings of the Regulatory Model can also be seen in other sectors, including social sectors like education (Mehta 2011, Tiwari 2005). In most of the cases, the regulations, based on the preventive strategies and the unrealistic ceiling on the pricing, do not necessarily enhance the capability of the target groups at the end.

Both the World Bank model and the Regulatory model based on the PPP are somewhat in tune with the proposal outlined in the CDP for Delhi, prepared under the JNNURM of the Government of India for a seven-year period. As a conceptual departure from the overwhelmingly welfare-state-driven Master Plan documents, the CDP of Delhi has essentially looked at the partnerships between the central government, the state government, urban local bodies, and financial institutions in improving basic services of water supply in the city (IL&FS EcoSmart Ltd 2006).[11]

Two key points emerge from discussions on these three broad models: one, the conceptual departure from the complete control of the "welfare state" in the provision of urban services, like water supply, and the other, the acceptance of the techniques and mechanisms of what Foucault (1975[1972], 1984) refers as the "management state," by entrusting the private agencies for the end-delivery, while regulating operations of the same. Both these positions are interconnected. Business and commercial interests, as urban geographer Swapna Banerjee-Guha (2000, 2002) perceives in the context of Mumbai as well, seem to drive the broader urban planning policy of the government of supporting the privatization of infrastructure. Now the question is: In this state of affairs, what are the possible structural approaches to be adopted for reducing inequity and injustice in service delivery and more specifically, in water supply? In place of the preoccupation for economic and demographic factors, equity concerns have become a line of argument, which tends to bring out *fairness issues* and recognize "studies of proportionality and balance in the distribution of resources" (Davidoff and Boyd 1983, Krumholz and Forester 1990). Noted urban planner, Fainstein (2000: 473), too, underlines that the equitable distribution, an important and necessary consideration in the urban planning discourse, should seek for "a more pluralistic, cooperative, and decentralized form of welfare provision than the state-centered model of bureaucratic welfare state."

The notion of the "welfare provision" appears to emerge out of a concern for achieving equity and justice by increasing required access of water supply to all sections of society. Empirical findings on the delivery of water supply, however, indicate that the existing government-controlled mechanism deviates from both the original ideological position and the policy intention of providing the *just* delivery. Such a mechanism eventually picks up the governmentality techniques of differential implementation of the policy and, in effect, often deprives the lower strata of society. The state government's relatively recent regulatory model, attempting to overcome broader territorial unevenness of the production and distribution of water in Delhi, addresses only one dimension of inequity. But, inequity out of differences in actual delivery across socioeconomic groups and spaces within each territory of the service zones (or command areas, or OZ) of the DJB remains unattended. It is interesting to see the possibilities of how the minimum access model and the pricing premiums, I have discussed, can be imbibed into larger concepts of the structural modification.

Options of Control and Flexibility

Neoliberal ideological constructs, influencing the Delhi Government's "reform" proposals of the World Bank model or the Regulatory Model of privatization of water, have been central to worldwide trends and policy discourses in the water sector reforms, especially since the 1990s (Budds 2004).[12] The haunting question remains, how can water supply be made equitable and efficient? Therefore, what is significant is the possible combination of the control of the state and the flexibility to be given to others. This brings the discourse of the Structure-Agency to the fore.

As against too much of "Statist" interventions of control or too much leniency of allowing flexibility by depending upon the private "actors," one would like to address the issue through a very basic understanding of the theory of Structuration proposed by sociologist Anthony Giddens and analyzed by others (Barley and Tolbert 1997, Colley 2001, Fuchs 2003, Giddens 1982, 1984). In the Structure-System-Structuration, Giddens (1982: 35) identifies Structure as "recursively organized rules and resources and it only exists as 'structural properties'", System as "reproduced relations between actors or collectivities, organized as regular social practices" and Structuration as "conditions governing the continuity or transformation of structures, and therefore the reproduction of systems." The central argument is:

> [N]either the subject (human agent) nor object ("society" or social institutions) should be regarded as having primacy. Each is constituted in and through recurrent practices. Explication of this relation thus comprises the case of an account of how it is that the Structuration (production and reproduction across time and space) of social practices take place. (Giddens 1982: 8)

Interestingly, Giddens (1982: 9, 1984) considers two components of human conduct: "capability" and "knowledgeability" (Barley and Tolbert 1997, Fuchs 2003). The first stands for "power" of the agent to act differently and the second indicates the information that the members of any society have about that society, helping them go on with their activities and life. Giddens's notions of capability can be connected with Sen's (2009) argument on justice and his concepts on knowledgeability can be understood as well-informed, decentralized, bottom-up mechanisms of identifying the scope and opportunity.[13]

Perhaps, both capability and knowledgeability can be applied to understand the context of water supply as well. The poorer strata of society, for example, would have to face more difficulty in meeting shortfalls of the institutionalized water supply than the richer one. The latter is likely to have the capability to afford alternative choices in this respect. Similarly, richer sections will have more options to augment poor quality of water by purchasing drinking water, by installing filtering gadgets, etc. Therefore, any such shortcoming in terms of capability parameters would affect the poorer section more and would then be considered unjust. This is, indeed, central to the argument put forward by Sen (2009). In fact, knowledgeability should help the community participation and advocacy in estimating water demands, controlling the supply at local levels, and sharing the cost of distribution of supply. Accountability in the water governance, too, is expected to increase in the process.

To extend this discussion a bit further, one may bring back the idea of the state and introduce the entity of enclave as a definitive physical and formal expression of the notions of community or of territory. Literally, an enclave is formed when a piece of land is surrounded by land owned by others. One would expect the state to increase the capability of the people and the enclave to have the knowledge and information of the people living within it.

Let me refer to a related discussion on water supply. Duncan Mara, an eminent expert in public health engineering, and Graham Albaster from United Nations Human Settlements Programme in a joint paper, suggest a "new paradigm" of providing water supply in urban areas to a group of individuals, termed as the "condominial supply," and not to individual households, referred to as the "conventional supply" (Mara and Albaster 2008: 119). With the help of the case study of water supply in Parauapebas town in the northern Brazilian state of Parà, they show that the cost per connection including exaction and piping of the condominial supply is about one-fourth of that of the conventional supply (ibid.). Subsequently, they suggest a model of water supply cooperatives in urban areas to be formed by groups of households for "condominial supply" situation, and propose three levels of cooperatives based on socioeconomic groups: "Stand-pipe Co-operatives (i.e., one or two stand-pipes per group of member households), Yard-tap Co-operatives (i.e., one tap per member household) and In-house Multiple-tap Co-operatives (i.e., individual full household connections for non-poor household groups)" (ibid.: 121).[14] This model of co-operatives essentially looks for a cheaper alternative and decentralized model for the provision of basic services, whereas the ownership or operational control of the service provision is not the central issue. In Delhi too, there exists a policy of "social supply" (sic.) by the state by providing stand-pipes in slum areas. But in reality, conditions of such supply, as mentioned earlier, result in the accrual of the "poverty premium" on the slum-dwellers to compensate for the shortfalls of water supply. Thus, objectives of the state are contradicted.

The Gurgaon example is also worth a mention in this discussion on the private supply mechanisms (Rao 1993). Gurgaon, a city to the south of Delhi in the state of Haryana within the National Capital Region (NCR), has widely been understood as an outcome of the neoliberal policies and the improved global networking (Biswas 2007, Biswas 2002, 2004[2002]). The city is popularly identified with the multinational corporate offices and high-end private housing enclaves. In Gurgaon, where large chunks of land-holding are under the private developers, water supply provision is often found to be a working combination of the two-tier state–enclave model (Figures 9.4 and 9.5).

Private developers are involved in installing the "internal infrastructure" such as pumping station, overhead tanks, water pipeline networks etc. within the enclave and the "external infrastructure" of providing the trunk services is undertaken by the government agency, like the Haryana Urban Development Authority (HUDA) (Rao 1993). The operation and maintenance of water supply in many high-end residential enclaves is handled at the enclave level (by the private developer group who has created it) and a lump-sum amount is charged as a part of the regular maintenance fees to individual households for their unmetered water supply. In turn, the developer group for that particular colony pays to the government agency (the state) for the total water allocated.

The moot point in discussing the notion of the "condominial" model of water supply suggested by Mara and Albaster (2008) and the Gurgaon situation is to emphasize the collaborative possibility between the state and the enclave (the state–enclave model). Interestingly, the emergence of the postmodern, post-informational cities, like Gurgaon, renders an opportunity for the state–enclave model. The state is expected to augment the capability of the people and to bring the resource of water up to the enclave. The enclave, with the knowledge and information of the people living within it, is to work out internal arrangements for supplying or delivering that water.

But post-independence planning of Delhi was conceived out of the welfare-state-driven policy. The state undertook the acquisition

Figure 9.4:
Enclaves in Gurgaon in 2004

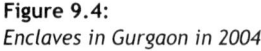

Source: Adapted from SSAA (2004).[15]
Note: Different shades indicate land pockets in the newer area of Gurgaon controlled by different private developers and HUDA around 2004. Land parcels were acquired and consolidated to form estates. Many enclaves also existed even within one estate. In Gurgaon land-holding conditions undergo continuous and rapid change. Today, one may see changes in the above land ownership map.

and redistribution of land in the master-planned Delhi, which, unlike Gurgaon, did not have the predominance of the enclaves owned and developed by private developers. Instead, the planning of Delhi intended to achieve a balanced distribution of socioeconomic groups over spaces following certain hierarchical arrangements. Such a chimerical objective driven by the utopia of the modern city has not been realized. The

Figure 9.5:
State—Enclave Model: Gurgaon Situation

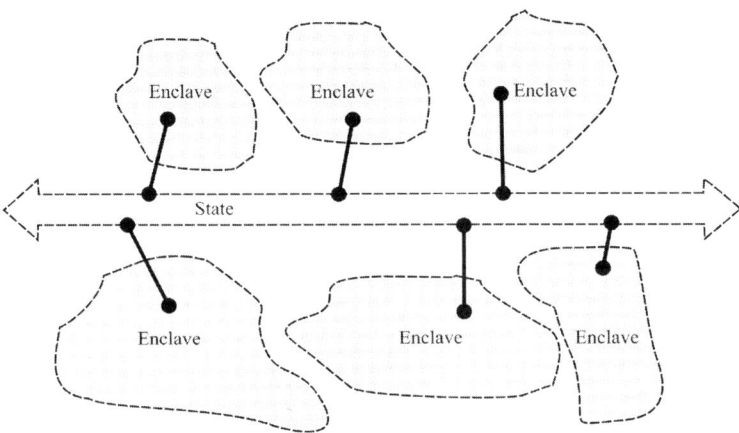

Source: Author.

unfulfilled utopia, which many refers as dystopia, is being constantly reminded by the presence of slums and unauthorized colonies as well as by inadequate provision of urban services, like water supply, among many such instances.

As the piped network is in place in most of the areas in Delhi, a three-level structure of the state–territory–enclave may be possible for the distribution, operation, and maintenance of water supply (Figure 9.6). In this model, an intermediate Local Service Provider for Water Supply, a private agency or a consumer cooperative or a public agency, can be introduced between the state and the enclave. In place of the service zones of the DJB, the territory of operation for such a service provider can be the political boundary, for example, the AC. When the lower supply is observed in the lower strata even within the same DJB service zone, one becomes apprehensive about the efficacy of such zoning of supposedly technical nature. Political accountability, on the other side, in the operation and maintenance of water supply is expected to be higher within such political spaces even though no clear-cut expositions of the theoretical models, like the *vote-bank politics* or *patron–client relationship*, are observed in the empirical research on the political funding.

Figure 9.6:
State—Territory—Enclave Model

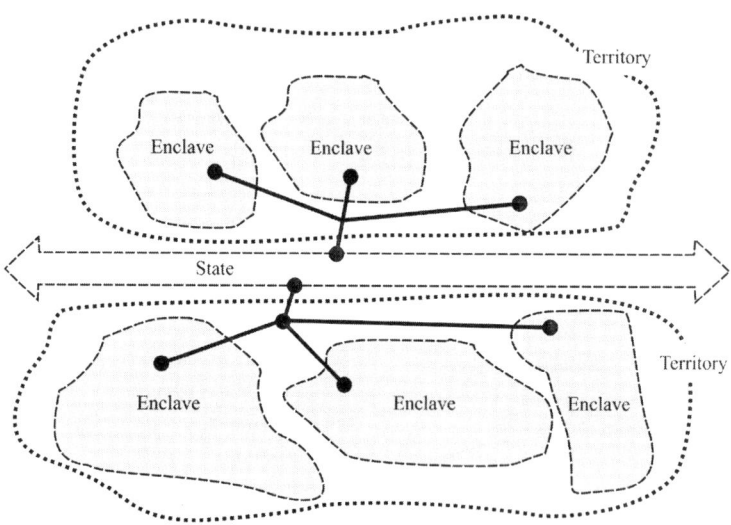

Source: Author.

The complexity of land acquisition by the state and the encouragement given to the private sector to accumulate land from the market have been witnessed for some time in India. In the long run, this process will catalyze the piecemeal and ad hoc urban growth with an assembly of enclaves. Such urban expansions tend to operate within the governmentality of multiple conditions out of unpredictable ownerships, plot-sizes, and land uses resulting in the creation of enclaves with differential accessibility. Thus, the state (the structure) and the autonomous enclaves (the agency) exist in dependency with each other where one enclave seldom has any relationship with the other.

There are subtle operational differences in the service delivery between the state–enclave and the state–territory–enclave models.

First, the state–enclave model is more effective for three co-operative situations proposed by Mara and Albaster (2008). It tends to depend more on public agencies for the supply up to the threshold of enclaves (or condominiums). The state–territory–enclave model, on the contrary, can accommodate a private service provider by giving them a territory of operation and maintenance which, in turn, may work with the co-operatives at the enclave level.

Second, in the first model, private local supply providers may not find sufficient encouragement to supply water to poorer enclaves because of less financial return out of the pricing premium strategy proposed earlier. In the second model, the private provider is supposed to control a larger territory, for example an AC, and will supply water for a wider range of the strata. In such situations, the service provider may have to cater to the lower strata of society, having a premium charge for per unit of optimal water supply much lower than that for higher strata. Even then, the financial return will get averaged out due to the presence of different strata within its territory.

Third, the state–enclave model seems more suitable for newer development in the city. The state will then develop the infrastructure from the scratch and extend it till the enclave level. The enclave will have its autonomous system to be connected with state infrastructure. But with the infrastructure already in place, the advantage of the cost reduction during initial installation at the enclave level does not count. For existing infrastructure, this model can only be used for operation and maintenance for which it has to rely on the state's performance. In an existing urban setting, the state–territory–enclave model creates an operational space for an intermediate private water supply provider having certain territory of control.

In these models, the state signifies the structural intervention of sourcing, treatment and major supply of water, the territory, the intermediate tasks of operation and maintenance, and enclave, the multiple information and operations at the actual delivery end.

Now, one is curious to see if micro components of equity measures, such as the optimal supply, the pricing premiums, and water demand calculations, which I have suggested a little while earlier, fit into such models of control and flexibility in water supply.

For a state–enclave situation, the two-slab supply of the *minimum access model* may ensure the complete coverage of the minimum supply stipulated by the state across all strata after which, the optimum supply will be provided in a differential scale open to the market condition and may be higher for higher strata. The *pricing premium*, a premium for the people from higher strata to pay more per unit of supplied water, too, may tempt private supply provider to give more optimal supply for the higher strata. However, the consumption of water, even in the higher strata, cannot be unlimited. The control of the state may intervene here through the regulatory guidelines to ensure different ranges of optimum supply across various socioeconomic strata. This range of optimum supply may be governed by the participatory, bottom-up approaches of the community-based methods of demand calculation at the enclave level. The co-operatives or the RWAs shall be a necessity for such mechanisms. A parallel to this demand calculation process would be the self-assessment of property tax in Delhi. Instead of the method or technique to assess the actual delivery at the ends, here, the capability of the users are estimated by themselves drawing from the knowledgeability of their own way of living. Perhaps, Fainstein's (2000: 473) notion of "pluralistic, cooperative, and decentralized" form of welfare provision, one may expect, would then be addressed.

On the one hand, these propositions may also be helped and simultaneously checked by the efforts of the "deep democracy" which, as eminent sociocultural commentator Arjun Appadurai (2001: 25–26) suggests, works within the "privatization of the state" through diverse mechanisms. These are the "new ways of claiming space and voice" at the *grassroots* level in poorer cities by poorer populations and by non-state actors, such as the advocacy groups, citizen's associations, and NGO, etc. On the other hand, the state as the service provider or as the regulatory agency has to play a significant role despite activities of the grassroots level organizations.

Jurisdictional Overlaps and the Dominance of Political Territories

The findings of the empirical study have shown that the difference in water supply in the colonies is not so significant across different DJB service zones covered under the first level of the DJB UGR. Such observations are made when DJB zones are seen in combination with the political and planning jurisdictions represented by the AC and the planning zone, respectively. Interestingly, significant effect of the political territory on water supply in residential colonies points at the importance of such a relationship, which otherwise may not be so apparent. Governmentality forms the wider theoretical explanation of variable practices of water supply that makes the delivery drift away from the policy of planning toward the sphere of politics (Chatterjee 1994, 2004, Rancière 1999). Territorial jurisdictions of the AC, also, have significant effect on the spending of the MLALAD funds through government agencies, such as the DJB and the MCD.

Since the delivery of urban services seems to be under the influence of politics and the political discourse in the representative democracy hinges around the control on the power of the state, it is necessary to bring political responsibility and accountability at the focus of urban governance. One simple technique would be to merge multiple territories of jurisdictions, like the delineation of the operation and maintenance zone of water supply, with the political territory. The political jurisdiction is the ideological space of equality, where the democratic right of voting is exercised.

On a rather general note, this line of interpretation may also be extended to the alteration of planning zones. The policy of planning, notionally and structurally, is kept independent of politics and that is the essence of both the operational and the territorial autonomy of planning. Such dichotomy of mutual exclusivity between the policy of planning and politics, Chatterjee (1994, 2004) underlines in the context of the post-independence India, appears to be founded on moral and social consensus. This is close to John Rawls' notion of justice upheld by the agreed upon conduct. However, in reality, planning as the delivery of policy of the state with certain rationalities of distribution tends to have techniques of governmentality embedded in it. I have made such explanations while analyzing the MPD (1962) earlier within the theoretical formulations of this work. Several objectives did not even get implemented or delivered on ground to realize what I refer as the *intended city* of the policy of planning. To reiterate, planning gets coerced by politics though the governmentality techniques of manipulation of policies and multiple strategies of implementation with inconsistent results and implications. Such assumptions call for the review of the overriding nature of planning territories and to avoid unnecessary territorial mismatch, it may be a better idea to make planning zones coincide with political territories.

However, these observations on territorial convergence, one expects, will instigate further explorations for more concrete conclusions.

Governmentality of multiple strategies are somewhat visible in the ambivalent nature of the political representatives, at least from the way the MLALAD spending has been made in the study areas within Delhi. Earlier in this work, I have elaborated two sides of the argument on possible arrangements of the patronage persuading the implementation of the policy (of distributions, etc.): on the one hand, the local-level political persons in alliance with the other powerful residents showing loyalty toward the rich (Harvey 1975[1973], Lasswell 1936, Logan and

Swanstrom 1990 [2005], Mollenkopf 1992, Molotch 1976, 1993) and, on the other hand, the *patron–client relationships* driven by the vote-bank politics between the politicians and the favored *political society* of the marginal and underprivileged population groups (Chakrabarty 1989, Chatterjee 2004: 40, Scott 1972).

However, contrary to these theoretical assumptions on the political patronage, the empirical analysis suggests that the spending of the discretionary political funds, like the MLALAD, over a period of five years in the study areas does not overwhelmingly favor either the rich or the poor. The MLALAD fund is the most relevant one for water supply in Delhi, and the fund utilization, perhaps unknowingly, tends to benefit the higher socioeconomic strata and spaces occupied by them. Notions of *vote bank* related spending or the *patron–client* relationship between the politicians and the lower socioeconomic strata seem limited in explaining general trends of the MLALAD distribution. Also, the agency-wise spending patterns are different. For example through DJB, more funds are allocated to the lower half of the socioeconomic spaces, whereas through the MCD, it is to the upper half. This situation can be seen as an example of multiple techniques and politics of governmentality.

Notes

1. This situation is far more relevant within the realms of techno-science debate especially regarding the access to technology, like information technology.
2. Harvey's (1975[1973]) reliance on the "urban ecology" model assumes that the spatial availability of man-made resources is dependent on location-based distributive decisions. Such decisions, the model explains, often influence variations across larger territory. Proximity and access to resources become significant interrelated aspects of distribution. On the other hand, Latour (1996), while indicating the properties of networks, explains notions of far/close as:

   All definitions in terms of surface and territories come from our reading of maps drawn and filled in by geographers. Out of geographers and geography, "in between" their own networks, there is no such a thing as a proximity or a distance which would not be defined by connectibility.

3. Instead of explaining how capability of poorer people seems constrained in easing out any difficulty, it may be apt to reiterate certain basic notions of capability here. An excerpt from one of Sen's (1979) earlier lectures, titled "Equality of What?", delivered at Stanford University, may bring out certain nuances in that direction:

> ... [The] notion of "basic capabilities": a person being able to do certain basic things. The ability to move about is the relevant one here, but one can consider others, e.g., the ability to meet one's nutritional requirements, the wherewithal to be clothed and sheltered, the power to participate in the social life of the community. The notion of urgency related to this is not fully captured by either utility or primary goods, or any combination of the two. . . . There is something still missing in the combined list of primary goods and utilities. If it is argued that resources should be devoted to remove or substantially reduce the handicap of the cripple despite there being no marginal utility argument (because it is expensive), despite there being no total utility argument (because he is so contented), and despite there being no primary goods deprivation (because he has the goods that others have), the case must rest on something else. I believe what is at issue is the interpretation of needs in the form of basic capabilities. This interpretation of needs and interests is often implicit in the demand for equality. This type of equality I shall call "basic capability equality."

4. Pricing for domestic use of water as per the DJB bill in 2011:

Monthly Consumption (in kiloliter)	Service Charge (in ₹)	Price per Kiloliter (in ₹)
0–10	50.00	2.00
10–20	100.00	3.00
20–30	150.00	15.00
>30	200.00	25.00

This pricing came into effect on January 1, 2010 and has been revised subsequently.

5. In general, importance of local level public participation for infrastructure governance is highlighted in a lot of works (Hoyt et al. 2005, Sahoo 2007, Savage and Dasgupta 2006, and many others). Hoyt et al. (2005: 10–14) discuss applications of GIS in public participation in order to use local knowledge "to improve public service delivery" and "to create more inclusive planning processes for water service delivery and infrastructure" in informal settlements in Delhi. From 2002 to 2003, National Institute of Urban Affairs took initiatives, namely the planning, learning, and action techniques, to collect ground level information in *informal* settlements in

eastern Delhi to document the growing demand–supply gap of water (Hoyt et al. 2005). However, most of these studies and initiatives tend to focus on settlements existing outside the planning document and there is a growing need to use similar efforts across all types of socioeconomic spaces in the city.

6. Organization of Economic Co-operations and Development (OECD) includes most of the European countries, Australia, Korea and Japan, Turkey, etc.

7. The World Bank's (2005) identified that the proposed project would be supported by a Specific Investment Lending operation. The Plan would include the first phase of water distribution and waste water collection improvement, trunk water and sewerage infrastructure improvement, organizational strengthening, urban water supply and sewerage services to the poor, and the roll-out plan.

8. Primary criticisms of the DJB's version of the proposed World Bank model were the following (Bhaduri and Kejriwal 2005):

   • inconsistent selection of consultants to prepare the project report on the reform plans,
   • involvement of multinational private companies and high-salary cost of their staff,
   • increase in tariff across all sections of consumers, and
   • lack of provision for extending the service to the urban poor.

9. As per an analysis of Delhi Water Supply and Sewerage Project, Parivartan (ca. 2005), a Delhi-based NGO, highlights that the total operational expenses were proposed in the project to be about 1,100 crores, whereas in reality, revenues collected for the year 2004–2005 were ₹400 crores. In other words, tariff will have to be increased almost three times to recover the increase in operational expenses and to pay huge salaries.

10. The Delhi Water Board Act 1998 (Delhi Act No. 4 of 1998) was passed by the Legislative Assembly of the NCT of Delhi. See for the complete act: http://delhijalboard.nic.in/djbdocs/about_us/act.htm (accessed April 14, 2011).

11. The CDP of Delhi, prepared by the private consulting firm Infrastructure Leasing and Financial Services Ecosmart, has identified three major issues regarding water supply situations in Delhi: wide variations in supply, UFW losses in transmission and distribution, and inadequate management framework (IL&FS Ecosmart 2006). Broad strategies envisioned for an efficient water supply have been the upgradation of the overall management of distribution, augmentation of supply, and strengthening of cost recovery mechanisms (ibid.).

12. "Neoliberal ideology holds that social functions are best managed through free market, and that economic development should be undertaken by the private sector, with the state playing a facilitating and regulatory role"

(Budds 2004: 322). The 1992 "Dublin Principles," too, underlines a similar neoliberal viewpoint. It recognizes "water as an economic good," which is to be managed not only to achieve its "efficient and equitable use" but also to encourage and protect water resources (WMO 1992, quoted in Budds 2004: 322).

13. Fuchs (2003), a social media theorist, observes relationships between Giddens's notion of knowledgeabilty and the information system. Giddens recognizes technology of information as an important factor in social modifications. On the one hand, people have the information about their settings and, on the other hand, the state has the information about its people and resources. Fuchs (2003) refers to a similarity between Giddens' under-standing of surveillance as the administrative power of storing and control of information, and Foucault's notion of the surveillance and disciplinary techniques. Recent controversies on "unique identity number" for citizens in India have discussed such kind of surveillance techniques of the state. However, as opposed to Foucault for whom surveillance is an instrument of power and control, Giddens considers it as a basic mechanism of integration.

14. The following outline of the tariff structure, indicated by Mara and Albaster (2008: 123), may explain the notion of these cooperatives better:

   - for stand-pipe co-operatives, the supply is not to be metered and a *nominal tariff* (a fixed monthly charge = a small percentage, say 1–2 percent, of the local minimum wage × number of member households) is charge-able to the co-operatives;
   - for yard-tap co-operatives, the supply is not to be metered and a *minimal tariff* (a fixed monthly charge = a bit larger percentage, say 3–5 percent, of the local minimum wage × number of member households) is charge-able to the co-operatives; and
   - for in-house multiple-tap co-operatives, the supply is to be metered and on the basis of a block tariff structure, the cooperatives are to be charged as per the actual consumption.

15. The drawing is from the unpublished research work done by the 4th year students in the Urban Housing Studio in 2004 in Sushant School of Art and Architecture, Gurgaon, India, under the faculty supervision of Suptendu P. Biswas, Vishal Aggarwal, Ashish Choudhury, and Ashish Bhalla.

10

Equity Mosaics and Assorted City

I would like to take this discussion toward a reading of the city. One significant formulation about the relationship between political community and people is Chatterjee's (2004) twofold construct of the "civil society" and the "political society" represented by citizens and populations, respectively. Chatterjee's concept, I have mentioned earlier, is conceptually akin to Balibar's (1994, 2005) idea of "property" and "community". The political society in Chatterjee's observations is grounded in an understanding of Gramscian thinking and subaltern criticisms on politics, and often stretches the limit of the legal framework, makes certain "para-legal" arrangements for its existence and, in doing so, finds itself closely linked with the political fraternity. Instead, the civil society, conceived in tune with the notion of citizenship, is supposed to enjoy the status available in a given democratic set-up (Chatterjee 2004). In fact, conceptions of dual urban existence are found in a rich body of work (Balibar 1994, 2005, Chatterjee 2004, Holston 1998, Holston and Appadurai 1996, Marshall 1950).

The majority of the city readings on Delhi, too, are based on, what I refer to as, the *two-city* notion. The central discourse is built upon the views on opposing realities of the planned and unplanned city. The planned city, the legally justified entity, intends to accommodate the *citizens* as "full members of a community" who supersede "local hierarchies, statuses, and privileges" and enjoy equal "rights and duties" (Holston 1998, Holston and Appadurai 1996: 187, Marshall 1950: 149). On the contrary, the unplanned city grown outside the planning provisions of the MPD is what the State considers the illegal urban existence. These mind-sets are predominant among the government authorities like the DDA, for whom planning is a legal tool. They often associate the "illegality", otherwise pervasive across all strata of society in Delhi, with the urban poor. However, several significant narratives recognize that the predicament of the urban poor is caused by the incomplete execution of the Master Plan and related bad governance (Baviskar 2002, Bhan 2009, Dupont 2004, 2008, Ghertner 2008, Kumar 2006, 2008, Menon Sen and Bhan 2008, Roy 2000, Sharan 2006, Sundaram 2010, Tiwari 2003, Verma 2002, Zérah 2000a, 2000b). Perhaps, the very existence of the urban poor in Delhi is hinged upon, to recap what Holston (2008: 253) identifies, a "hybrid mix of special treatment rights, text-based rights and contributor rights," which are, out of the "poverty or need," the "legal citizenship" and "his contribution to and work within the city," respectively (Bhan 2009: 133). This ambiguous status of urban living is seen as the compulsive condition of necessities forcing the "state of exception" (Agamben 2003[2005]). It is also well documented and understood in most of the works on Delhi, cited before, that there are wide gaps, in general, between these two conditions of living in accessing civic amenities and services.

However, this genre of discussions conceptualizes city as a place of binary opposing existences, legal–illegal, planned–unplanned, authorized–unauthorized, formal–informal, civil society–political society to name a few. In this work, I have attempted to go beyond such formulations to discuss that even within the legal (or planned) colonies, several gradations of socioeconomic stratification do exist and multiple differential delivery occur at the end. The central argument of this work hinges around the notion of multiplicity of urban living, exemplified by lower water supply to the lower strata of the society occupying the so-called *legal* spaces with supposedly full facilities. Similar scrutiny of para-legal situations of urban life would also have in it the ingrained shades of differences and variations. Such multiplicity of the urban living, I perceive, creates an *equity mosaic* in a city, representing various states of inequity and injustice. Often, issues of equity and justice are discussed as with reference to multiple conditions of inequity and injustice.

I find Ananya Roy's notion of "informality" worth a mention here. Roy, a scholar of international development and global urbanism, put forward the central argument in an article titled, "Why India Cannot Plan its Cities: Informality, Insurgence and the Idiom of Urbanization":

The two scenes of Indian urbanization described above when taken together can be seen to present an incontrovertible argument about the failure of planning in India: that informality and insurgence together undermine the possibilities of rational planning, and that therefore India cannot plan its cities. Against this narrative of failed planning, I present the argument that what is at work in the two scenes is an idiom of urbanization. This idiom is peculiar and particular to the Indian political economy and yet can be detected in many other contexts. While this idiom seems to be antithetical to planning, and indeed seems to be anti-planning, it can and must be understood as a planning regime. I also argue that the key feature of this idiom is informality. (Roy 2009a: 80)

Roy's earlier works (2003, 2005; with AlSayyad 2003) also indicate "informality" as a valid analytical framework to describe the urbanization process, especially in the "Third World" cities. Broad features of her argument, owing to its similar alignment as this work, need a definite mention here. Urban informality, as Roy (2003, 2005, 2009a, 2005; with AlSayyad 2003) proposes, has three key points: First, informality should not be mixed up with poverty; second, it is a system of deregulation with "calculated" actions; and third, it is frequently exercised from the top by the state. On the other hand, insurgent conditions prevailing and operating at the grassroots level and within the socio-spatial politics, too, do not necessarily bring about the *just city* (Roy 2009a). Her formulation of informality successfully moves beyond the dualities of several urban discourses and seems in allegiance with Foucault's idea of governmentality, but with a high reliance on the power of the state as opposed to notions of "counter-governmentality" from the bottom (Appadurai 2001, Chatterjee 2004). Roy (2009b), in fact, elsewhere also uses the term of "civic governmentality" and discusses Foucauldian construct of governmentality in support of the notion of urban informality. The assumption on the state's intentions to achieve an informal, rather, a flexible condition by degenerating urban planning, somehow, makes the state too ominous. If that is the case, it would be difficult to explain the state's continuous engagement with the urban planning in cities like Delhi, a model that is still being followed, perhaps inadvertently, in many smaller towns. Nevertheless, while applying the framework of informality to explain the urbanization, one may need to locate the discourse within a specific slice of time, because what is informal today becomes a precedent tomorrow to be followed or discarded as an existing rationality.

The IPG framework proposed in this work, however, helps in discussing the spatio-temporal narratives of the city. Notions of the politics of patronage, the selectivity of the State, and the state of exception taken from certain established, yet varied positions, I consider, can form a theoretical alliance to address the scaffolding (of the city and its urbanism) in transition from ideology to policy to governmentality in a recursive and replenishing way (Agamben 2003[2005], Chatterjee 2004, Harvey 1975[1973], Jessop 1990, Mollenkopf 1992, Molotch 1976, 1993).

Even in the State–enclave formulation, the State as a key entity returns to the main discourse. The city, like Delhi, is a vivid case of the production of the ideology of the *just city* as envisioned within the paradigm of the Welfare State, the policy of the *intended city* as stipulated by the urban planning, and the governmentality of the *real city* subversive of the first two, thereby continuously trying to maneuver those. And the city keeps on changing itself.

The *real city* of Delhi on ground *is a site of heterotopias*, which indeed, is *an assortment* full of inconsistent, contradictory, and parallel realities of what was intended by the Master Plan and built accordingly the legal and what was not illegal/para-legal.

I address inconsistent, contradictory, and parallel realities of the real city—*an assortment*, from the perspective that a contemporary city forms heterotopias. Foucault (1984[1967]) introduces heterotopias as "a single real place made up of several spaces, several sites that are incompatible." The notion of heterotopias or *other* spaces was initially discussed in a lecture in French, titled "Des Espace Autres," that Foucualt delivered in the French Architecture Research Institute in 1967. Much later in mid 1980s, his lecture was published in English and has gained prominence.

My first encounter with the notion was in Georges Teyssot's (1977) article, "Heterotopia and History of Spaces." My understanding of heterotopia as a possible spatial construct is informed by Foucault's own work along with many other writings that followed (Angermüller and Bunzmann 2000, Barnes 2004, Foucault 1984[1967], González 2004, Hetherington 1997, Mcleod 1996, Saldanha 2008, Selman 2008, Shane 2007[2005], 2008[2000], Stavrides 2007, Teyssot 1977). Here, the objective is not to explain in details various concepts of heterotopias that Foucault highlights in his essay, but to connect certain specific notions of heterotopias in tune with the overall thematic direction of this work.

The term, heterotopia, originally used in the field of anatomy, refers to "parts of the body that are either out of place, missing, extra, or, like tumors, alien" to the regular body (Hetherington 1997: 42). The concept of heterotopias generates divergent views. The notion seems to encompass "a collective experience of otherness" (Stavrides 2007) and is a "politicized" perception of formal and social *otherness* (Mcleod 1996). Heterotopias are "places of Otherness [or] spaces, whose existence sets up unsettling juxtapositions of incommensurate 'objects' which challenge the way we think, especially the way our thinking is ordered" (Hetherington 1997: 42). Heterotopias are also seen as "spaces of possibility" (Selman 2008) and "are always in the process of being made, ordering rather than order" (Barnes 2004: 576). "Heterotopias as sites of contestation" also seem to indicate "the cohabitation of different social classes or cultural groups" (González 2004: 175).

There are certain criticisms of the idea as well. There has been circumspection regarding the structuralist dominance in the concept of heterotopia and with a hidden overture of the "totalist" understanding of spaces, the idea appears to make an inadequate analysis of spatial differences (Saldanha 2008). David Harvey also describes heterotopia as "incommensurable spaces" and wonders "what is the critical, liberatory, and emancipatory point of that [heterotopia]?"(Harvey 1990: 48, 2000: 538). Critics, however, recognize the conceptualization of heterotopia as a commendable effort of coining a singular term for "spatial differentiation" (Saldanha 2008).

Urban designer, David Grahame Shane's work (2007[2005]), *Recombinant Urbanism: Conceptual modeling in Architecture, Urban Design and City Theory*, is significant in this regard, where he extensively illustrates various combinatory options involving three elements of contemporary urbanism: enclave, armature, and heterotopias. Heterotopia, for Shane, functions to retain "the city's stability as a self-organizing system" and is "a place that mixes the stasis of the enclave, with the flow of an armature, and in which the balance between these two systems is constantly changing" (ibid.: 231). Postmodern cities would no longer be without a "master plan or a master planner"; instead, with "multiple actors by a spaghetti tangle of relationships produces patches of only local order, and no obvious mechanism of overall combination" (Shane: 305–306). Earlier, the notion of the "collage city," too, hints at autonomous figurative patches constituting the city (Rowe and Koetter 2001[1983]). However, Grahame Shane's work, is, perhaps, the most exhaustive and recent account of urban design and city reading of postmodern cities of the globalization era. Expectedly, the notion of "recombinant urbanism" coalesce urban design discourses on form, space, and design. The notion is discussed with the contours of poststructuralist philosophies, and, may be seen as a lazy onlooker of the politics of the city. The role of the state is somewhat absent in the construct. The armature may not necessarily be an urban element, as elaborated in Shane's work, but a structural relationship in the form of the infrastructure of roads, networks of water and electricity, urban management, etc., that the state can bring in to initiate the urban process.

Various descriptions of heterotopias, as Foucault (1984[1967]) discusses in *Of Other Spaces: Utopias and Heterotopias*, are, indeed, imbued with the ambition to explain the world of spaces, often inadequately stretching the notion too far toward generalization. However, my interest, here, revolves around his principle of "heterotopology" that heterotopia is an assortment of "incompatible spaces" juxtaposed and combined together to constitute a "real place" (Foucault (1984[1967). In fact, I began this discussion with such an understanding.

In the contemporary city, one recognizes the reality of autonomy of differences as opposed to the homogeneity that urban planning often aspires to achieve, rather erroneously. To underline it further, one may recall Iris Marion Young's observations on "city life and difference":

> As a normative ideal, city life instantiates social relations of difference without exclusion. Different groups dwell in the city alongside one another, of necessity interacting in city spaces. If city politics is to be democratic and not dominated by the point of view of one group, it must be a politics that takes account of and provides voice for the different groups that dwell together in the city without forming a community. (Young 2011[1990]: 227)

Observations on spaces of inequitable distribution of services, if one may revisit, form a narrative of differential spaces. Even if these spaces are adjacent to each other, such a narrative may not be location-centric. Instead, it is very often related to the socioeconomic strata. These spaces, despite being at variance with one another, coexist with multiple access, opportunities, and capabilities to avail the resource on offer. I have illustrated such issues with respect to the delivery of water supply.

Nevertheless, the very intention of planning and building the city might not be to produce such discordance. Over time, ruptures have been developed in the original essence of equity and social justice in the city. Eventually, at a given time, for example, now, the city is full with spaces of multiple conditions of equity and justice. In a spatial sense, pockets (or enclaves) of spaces with several grades (or degrees) of deviations from the original conceptions (of the ideology and the policy of planning) of the city can be found within a slice of time. Reasons for such deviations could be many. But, when the state in control of things succumbs to inconsistency and subsequently adopts it, certain exceptions are made in creating particular situation/s or selecting specific space/s or choosing certain group/s of people. As a result, there is a breakaway from the laid down policy framework and such incidents tend to happen time and again, case by case, by slowly dismantling the instrument. Urban planning is one of such instruments. Thus, conditions of governmentality are produced by the politics of distribution creating the difference in the city and vice versa.

The curious incursion of politics in this discussion occurs mainly because of the provenience of heterogeneity that politics in democracy has by bringing together diverse arguments at a given time.[1] In a wider sense, governmentality decisions are convinced more often than not by politics due to changing ground realities and democracy at the bottom, on the one hand, and due to the top-down wisdom (sic.) and the power of the state apparatus, on the other hand (Appadurai 2002, Gandy 2004). All of these lead to the cases, rather spaces, of exception because of the favor or neglect bestowed upon those and, thus, spatially and socially disaggregated enclaves are formed. Singur in West Bengal with the politics of land, paradoxes between industry and agriculture, and subsequent efforts of making special legal framework for the redistribution of acquired land is one such current example of exceptions and an enclave as well.

Enclaves, I am quite convinced, are produced by exceptions. These are exceptions to the *main* parts of the city or the mainstream policy or the *planned* city or dominant spaces and many such entities. Following the construct of exception-enclave, heterotopias as the collection of enclaves with inherent incompatibilities and similarities are formed in the city (Figure 10.1). Such conditions, which are not necessarily a terrible spatial segregation but practices of scattered collections of different forms of urban living, can, perhaps, create an assorted city (Figure 10.2).

Figure 10.1:
Physical Manifestation of an Assorted City in the Making

Source: SSAA (2011).[2]
Note: This example in a newer part of Gurgaon is a snap-shot taken in the beginning of 2011. Enclaves were being assembled not as dissimilar, but as differential identities and spaces.

Figure 10.2:
Recursive Relationships in the Assorted City

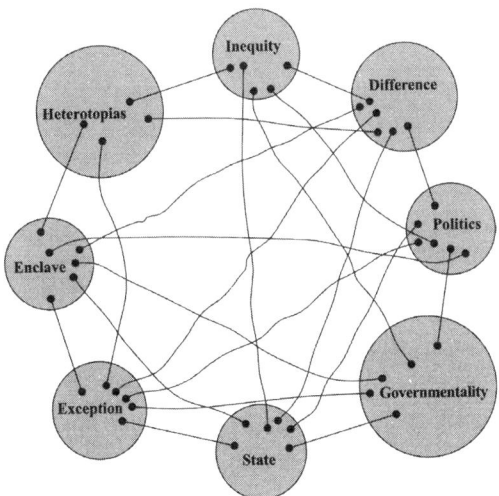

Source: Author.

Some initial thoughts on the notion of the assorted city may be shared here, which, indeed, requires more deliberations in time to come. But, its potential of evolving into an analogical framework to understand contemporary cities seems pertinent.

> Etymologically, assort (v.): late 15c., "to distribute into groups," from M. Fr. *assortir*(15c.) from O. Fr. *assorter* "to assort, match, from *a*- "to" + *sorte* "kind".
> Assortment: 1610s, "action of assorting," from assort + -ment. Sense of "group of things of the same sort" is attested from 1759; that of "group of things whether the same sort or not" from 1791.[3]

> *Assorted (adj.): varied, miscellaneous; sorted out, classified; matched, suited*
>
> (Landau 2002 edition).

Let me take the analogy of a box of assorted chocolate to explain the primary essence of the notion.

Different chocolates may have been put together in the box with a purpose of giving different tastes (the reason of richness), or with an intention to balance the cost of expensive chocolates in the basket (the reason of affordability), or due to unavailability of the number of a particular kind of chocolate required to fill up the box (the reason of scarcity). So, there may be any of these differing reasons behind the similar assortment. For the retailer, too, same box would attract more customers because it can address their varied needs. Researches also show that it may be optimal to use an "unprofitable" product in the assortment to avoid consumer search elsewhere (Cachon et al. 2002). My notion of the assorted city is somewhat loosely based on such basic observations and takes clues from few works on the assortment planning used in retail management (Cachon and Gürhan 2007, Cachon et al. 2002). Assortment planning is essentially a process to select the products to maximize sales and profit for a specific time period. One may notice similarities in the way private real estate enclaves are getting built on the outskirts of major cities in India. "Build that sells" is perhaps the only mantra of all such real estate initiatives. In fact, that is the growth-trend in Indian cities when the state has almost withdrawn itself from actively building the city.

Interestingly, the second and the third reasons of assortment may contribute to the distribution of basic services, infrastructure, and socioeconomic groups. The first may be applied in enriching the quality of urban life by using the mix of uses and people. In the assortment, an unprofitable product may well be found useful; similarly, at an enclave level, infrastructure and amenities with lower direct financial gains can also attract people to reside. The mix of housing for different social groups, even with less return per square foot of area, may eventually make better business sense. For a neo-liberal market-driven situation, such a mix of socioeconomic groups is expected to widen the cross section of users. When people can avail of such options in the city, the state (the structure) needs to lay down certain framework for possible arrangements. In the postmodern, "informational city," the state steps in for structural exigencies, recommends certain land uses and land holding guidelines with an intention to bring certain order in the pattern of urbanization. However, inept performance of the state in cities like Gurgaon often undermines this mechanism (Figure 10.1). The autonomous assorted arrangements of enclaves in terms of people, land uses, ownerships, urban form, and spaces make dissimilar juxtapositions inevitable. Even if the resultant urbanism appears random, it has an inbuilt logic of the market mechanism, which simultaneously triggers off corresponding politics of continuous change in land ownership, amalgamation and consolidation of land parcels, and conversion of land uses.

A transitional city (from the welfare state to the neo-liberal one) like Delhi, in contrast, is an example of unavoidable disintegration of the top-down planning display though the governmentality techniques of modifications, both from the state's end and from the ground, for the reasons discussed at length at various stages in this work. An overtly ordered system, thus, becomes a haphazard assortment of enclaves. In both Delhi and Gurgaon representing two divergent types of urbanization, the role of the state once again becomes important for its regulatory capacity in ascertaining urban design and planning guidelines within which the enclaves would make specific arrangements of people and spaces. An inherent structure and the casualness of agencies in the notion of the assorted city may not be overlooked for the time being.

$$* \quad * \quad * \quad * \quad * \quad *$$

This work may disappoint many who expect a singular ground-breaking idea to emerge. Instead, attempts are initiated to study and explain smaller, interconnected phenomena. At the theoretical level, the construction of the IPG framework, the formation of necessary and tactical theoretical alliances between the Marxist and the poststructuralist positions, and at an operative level, the use of property tax as an indicator for the socioeconomic spaces and the exposition of multiple shades of equity and justice in the city are some of the annotations this work produces. Conceptualization of the two-slab model for water supply, the state–territory–enclave formulation and possibilities of city reading are other related observations. Each one of these and other explanations in this research, to me, do not necessarily induce an impact that would turn the world on its heads. These arguments in themselves are like *weak ties* or threads which when interwoven carry certain intrinsic tenacity. The fishing net is the first example that comes to my mind.

Notes

1. Heterogeneity is what makes politics distinct from "juridical and commercial exchange" as well as from "religion and war" (Rancière 1999: 50). Rancière's notion (ibid.) that "democracy is the institution of politics itself," underlines the dimension of politics in this book.
2. The drawing is from the unpublished research work done by the 4th year students in the Urban Design Studio in 2011, titled "Nodes, Networks and Tissues," in Sushant School of Art and Architecture, Gurgaon, India, under the faculty supervision of Suptendu P. Biswas, Anjali Krishan, Leon Morenas, and Mitu Mathur.
3. Online Etymology Dictionary, available at http://www.etymonline.com (accessed on June 3, 2011).

Select Bibliography

Abercrombie, N., S. Hill, and B.S. Turner. 1994[1983]. "Determinacy and Indeterminacy in the Theory of Ideology." In *Mapping Ideology*, edited by S. Zizek (published in 1994), 152–166. London/New York: Verso.

Agamben, G. 2005[2003]. *State of Exception*. Translated from Italian by K. Attell. Chicago and London: The University of Chicago Press.

Alexander, C. 1965. "City Is Not a Tree." *Architectural Forum*, 122 (1) [Reproduced in G. Bell and J. Thywitt (eds). 1972. *Human Identity in Urban Environment*. Hammondsworth: Penguin, 401–426].

Ali, S. 1995. *Environment and Resettlement Colonies of Delhi*. New Delhi: Har Anand Publications.

Althusser, L. 1970. "Ideology and Ideological State Apparatuses (notes towards an investigation)." In *Mapping Ideology*, edited by S. Zizek (published in 1994), 100–140. London/New York: Verso.

Angermüller, J. and K. Bunzmann. 2000. "Hybrid Spaces—Theory and Beyond: An Introduction." In *Hybrid Spaces, Theory, Culture, Economy*, edited by J. Angermüller et al., 1–12. New York/Münster: Transaction/LIT.

Appadurai, A. 2001. "Deep Democracy: Urban Governmentality and the Horizon of Politics." In *Environment and Urbanization*, 13–23. Retrieved from http://eau.sagepub.com/content/13/2/23 (accessed 30 February 2011). doi:10.1177/095624780101300203.

Bhaduri, A. and A. Kejriwal. 2005. "Urban Water Supply, Reforming the Reformers." *Economic and Political Weekly*, 40 (53), 5543–5545.

Bahl, R.W. and J.F. Linn. 1992. *Urban Public Finance in Developing Countries*. New York: The World Bank/Oxford University Press.

Bakker, K. 2003a, December. "Archipelagos and Networks: Urbanization and Water Privatization in the South." *The Geographical Journal*, 169 (4), 328–341.

———. 2003b. *Good Governance in Municipal Restructuring of Water Supply: A Handbook*. Federation of Canadian Municipalities and Program on Water Issues at University of Toronto's Munk Centre for International Studies, July. Ontario

Bakker, K. 2005. "Neoliberalizing Nature? Market Environmentalism in Water Supply in England and Wales." *Annals of the Association of American Geographers*, 95 (3), 542–565.

Balibar, E. 1991. "The Nation Form: History and Ideology." In *Race, Nation, Class: Ambiguous Identities*, edited by E. Balibar, 86–106. Retrieved from www.univie.ac.at (accessed January 1, 2010).

———. 1992. "Foucault and Marx: The Question of Nominalism." In *Michel Foucault Philosopher*, edited by T.J. Armstrong, 38–56. New York: Routledge.

———. 1994. *Masses, Classes, Ideas: Studies on Politics and Philosophy Before and After Marx*. New York: Routledge.

———. 2005. "Difference, Otherness, Exclusion." *Parallax*, 11 (1), 19–34.

Banerjee, B. 2002. "Security of Land Tenure in Indian Cities." In *Holding their Ground: Secure Land Tenure for the Urban Poor in Developing Countries*, edited by A. Durand-Lasserve and L. Royston, 37–58. London: Earthscan Publications.

Banerjee-Guha, S. 2000. *Ideology of Urban Restructuring in Mumbai: Serving the International Capitalist Agenda*. Paper presented at the Second International Critical Geography Conference: For alternative 21st century geographies, Taegu University, Taegu, Korea, August 9–13. Retrieved from http://econgeog.misc.hit-u.ac.jp/icgg/intl_mtgs/SBGuha.pdf reduce space (accessed October 12, 2010).

———. 2002, January 12. "Shifting Cities: Urban Restructuring in Mumbai." *Economic and Political Weekly*, 37 (2), 121–128.

Barley, S.R. and P.S. Tolbert. 1997. *Institutionalization and Structuration: Studying the Links Between Action and Institution*. ILR Collection, Articles and Chapters (Paper No. 130). Retrieved from http://digitalcommons.ilr.cornell.edu/articles/130 (accessed June 20, 2011).

Barnes, T.J. 2004. "Placing Ideas: Genius Loci, Heterotopia and Geography's Quantitative Revolution." *Progress in Human Geography*, 28 (5), 565–595.

Barnett, C., N. Clarke, P. Cloke, and A. Malpass. 2008. "The Elusive Subjects of Neoliberalism: Beyond the Analytics of Governmentality." *Cultural Studies*, 22 (5), 624–653.

Barr, N.A. 1992. *Economic Theory and the Welfare State: A survey and Interpretation*. London: LSE Research. Retrieved from http://eprints.lse.ac.uk/archive/00000279 (accessed June 30, 2005).

Barraqué, B. 2004. "Not Too Much but Not Too Little: The Sustainability of Urban Water Services in New York, old Paris, and New Delhi." In *Sustaining Urban Networks: The Social Diffusion of Large Technical Systems*, edited by O. Coutard, R.E. Hanley, and R. Zimmerman, 188–202. London: Routledge.

Batley, R. 1996. "Public–Private Relationships and Performance in Service Provision." *Urban Studies*, 33 (4–5), 723–751.

Baviskar, A. 2002. "The Politics of the City." *Seminar*, 516 (August), 40–42.

———. 2006. "Demolishing Delhi: World Class City in the Making." *Mute Magazine*, September 5. Retrieved from http://www.metamute.org (accessed June 27, 2008).

Bhan, G. 2009. "This Is No Longer the City I Once Knew: Evictions, the Urban Poor and the Right to the City in Millennial Delhi." *Environment and Urbanization*, 21 (1), 127–142.

Biswas, S.P. 2002. "Gurgaon: A Mega-Corporate Park." *Architecture+Design*, XIX (6), November–December.

———. 2004[2002]. "The Ghost of Modernity: Architecture and Urbanism in the Era of Globalization." Reprinted in *Architecture+Design*, XXI (12), (December) [Original publication: *Architecture+Design* (November–December 2002, XIX [6]).

———. (ed.) 2007. *Dialogues: A Symposium on the Shaping of Gurgaon*. Gurgaon: Sushant School of Art and Architecture.

Bordreau, J.A. 2003, December. "Questioning the Use of 'Local Democracy' as a Discursive Strategy for Political Mobilization in Los Angeles, Montreal and Toronto." *International Journal of Urban and Regional Research*, 27 (4), 793–810.

Bourdieu, P. and T. Eagleton. 1991[1994]. "Doxa and Common Life: An Interview." In *Mapping Ideology*, edited by S. Zizek, 265–277. London and New York: Verso.

Brenner, N. 2000, June. "The Urban Question as a Scale Question: Reflections on Henri Lefebvre, Urban Theory and Politics of Scale." *International Journal of Urban and Regional Research*, 24 (2), 361–378.

———. 2004. *New State Spaces: Urban Governance and the Rescaling of Statehood*. Oxford: Oxford University Press.

Bridge, G. and S. Watson (eds.). 2002. *The Blackwell City Reader*. MA: Blackwell Publishing Ltd.

Bromberg, A., D. Gregory, G.D. Morrow, and D. Pfeiffer. 2007. "Editorial Note: Why Spatial Justice?" *Critical Planning*, 14 (Summer), 1–4.

Budds, J. 2004. "Power, Nature and Neoliberalism: The Political Ecology of Water in Chile." *Singapore Journal of Tropical Geography*, 25 (3), 322–342.

Burchell, G., C. Gordon, and P. Miller (eds). 1991. *The Foucault Effect: Studies in Governmentality*. Chicago: The University of Chicago Press.

Cachon, G.P., C. Terwiesch, and Yi Xu. 2002. "Retail Assortment Planning in the Presence of Consumer Search." Retrieved from http://opim-sun.wharton. upenn.edu/~cachon/pdf/searchv1.pdf (accessed June 10, 2011).

Cachon, G.P. and A. Gürhan Kök. 2007, June. "Category Management and Coordination in Retail Assortment Planning in the Presence of Basket Shopping Consumers." *Management Science*, 53 (6), 934–951. doi:10.1287/mnsc.1060.0661.

Castells, M. 1977. *The Urban Question*. Cambridge, MA: MIT Press.

Chakrabarty, D. 1989. *Rethinking Working Class History: Bengal 1890–1940*. New Delhi: Oxford University Press. Retrieved from http://www.books. google.com (accessed October 15, 2010).

Chatterjee, P. 1994. *The Nation and Its Fragments: Colonial and Post-colonial Histories*. Princeton, NJ: Princeton University Press.

———. 1997. *Our Modernity.* Rotterdam/Dakar: Sephis Codesria.

———. 2004. *The Politics of the Governed.* New York: Columbia University Press.

Colley, H. 2001. "Problems with 'Bridging the Gap': The Reversal of Structure and Agency in Addressing Social Exclusion." *Critical Social Policy*, 21 (3), 337–361.

Colquohoun, A. 1991[1989]. *Modernity and Classical Tradition: Architectural Essays 1980–1987*. MA/London: The MIT Press.

Corbridge, S., G. Williams, M. Srivastava, and R. Veron. 2005. *Seeing the State: Governance and Governmentality in India.* Cambridge, UK: Cambridge University Press.

CPHEEO (Central Public Health and Environmental Engineering Organization). 1999, May. *Manual on Water Supply and Treatment* (3rd edition). New Delhi: Ministry of Urban Development, Government of India.

Curtis, B. 2002. "Foucault on Governmentality and Population: The Impossible Discovery." *Canadian Journal of Sociology*, 27 (4, Fall), 505–533. Retrieved from www.elseminario.com.ar (accessed January 3, 2010).

Davidoff, P. and L. Boyd. 1983. "Peace and Justice in Planning Education." *Journal of Planning Education and Research*, 3, 54.

DDA (Delhi Development Authority). ca. 1961. *Draft Master Plan for Delhi 1962*. New Delhi: Delhi Development Authority.

———. 1962. *Master Plan for Delhi, 1962*. New Delhi: Delhi Development Authority.

———. 1990. *Master Plan for Delhi: Perspective 2001*. Gazette of India, Extra ordinary, Part II, Section 3, Sub-section (ii) vide S.O.606 (E), August 1, 1990, Delhi.

———. 1998, June 5. *Zonal Development Plan for Zone F (South Delhi-I)*. Authenticated vide No. K-13011/2/94-DDIB. Retrieved from http://www. dda.org.in/planning/docs/Zone%20F.pdf (accessed March 21, 2009).

———. ca. 2006. "Introduction." *MPD 2021: Vision Plan*. Retrieved from http://www.dda.org.in/ planning/docs/001-introduction.pdf (accessed April 12, 2007).

———. 2007a. *Master Plan for Delhi 2021*. New Delhi: JBA Publishers.

———. 2007b. *Draft Zonal Development Plan of Zone "F" 2021 (South Delhi-I)*. Approved by DDA vide Resolution No. 86/2007 dated 3.10.07 for inviting objections/suggestions.

Deakin, E. 1999. "Social Equity in Planning." *Berkeley Planning Journal*, 13, 1–5.

Dean, M. 1999. *Governmentality: Power and Role in Modern Society.* London: SAGE Publications.

Dear, M. and A.J. Scott (eds.). 1981. *Urbanisation and Urban Planning in Capitalist Society.* London: Methuen.

Department of Urban Development. 2009. *Guidelines for Member of Legislative Assembly Local Area Development Scheme (MLALADS) Delhi-2009.* Government of Delhi. Retrieved from http://www.delhi.gov.in/wps/wcm/connect/doit_udd/Urban+Development/Our+Services/MLALAD+Scheme/ (accessed March 6, 2010).

Dewan, H. 2004. "State Called Delhi, Are We Close Yet?" *The Times of India,* May 31, p. 2.

Dhar Chakrabarti, P.G. 2001. "Delhi's Ongoing Debate on Informal Settlements and Work Places—Issues of Environmental Jurisprudence." *International Workshop on "Coping with Informality and Illegality in Human Settlements in Developing Cities."* Network Association of European Researchers on Urbanization in the South (N-AERUS) and European Science Foundation (ESF), Leuven and Brussels, Belgium, May 23–26. Retrieved from http://www.ucl.ac.uk/dpu-projects/drivers_urb_change/urb_infrastructure/pdf_land%20tenure/NAERUS_ESF_chakrabarti_Delhi_Informal_Settlements.pdf (accessed August 4, 2008).

Dhawan, H.K. (ed.) 2004. *A Practical Guide to the Unit Area Method of Property Tax* (Abridged version 2004–2005). New Delhi: Allied Publishers Pvt. Ltd.

Dikeç, M. 2005. "Space, Politics and the Political." *Environment and Planning D: Society and Space,* 23 (2), 171–188.

Dimri, A. and A. Sharma. 2006. *Living on the Edge and Paying for It: A Study in Sanjay Colony, Okhla Phase II.* (Delhi CCS Working Paper No. 148), Summer Research Internship Program. New Delhi: Centre for Civil Society. Retrieved from www.ccs.in (accessed June 10, 2009).

DJB (Delhi Jal Board). 2004. *Delhi Water Supply and Sewerage Sector Reform Project.* New Delhi: Delhi Jal Board. Retrieved from http://www.delhijalboard.nic.in/djbdocs/reform_project/docs/docs/doc_project_prep_docs/introduction/DJB-ReformProject%20-%20Final.htm (accessed September 14, 2009).

———. 2004–2005. *Water Supply Schemes under Water Bulk for the Year 2004–05.* New Delhi: Delhi Jal Board.

———. 2006–2007. *Annual Plan for Water Supply and Sanitation 2006–07.* New Delhi: Delhi Jal Board. Retrieved from http//delhiplanning.nic.in/reports/PDF/Water.pdf (accessed June 30, 2006).

Donzelot, J. and C. Gordon. 2008. "Governing Liberal Societies—The Foucault Effect in the English-speaking World." *Foucault Studies,* 5, 48–62.

Dupont, V. 2004. "Socio-spatial Differentiation and Residential Segregation in Delhi: A Question of Scale?" *Geoforum,* 35 (2), 157–175.

Dupont, V. 2008. "Slum Demolitions in Delhi since the 1990s: An Appraisal." *Economic and Political Weekly*, July 12, 43 (28), 79–87.

Dutta, V., S. Chander, and L. Srivastava. 2005. "Public Support for Water Supply Improvements: Empirical Evidence from Unplanned Settlements of Delhi, India." *The Journal of Environment and Development*, 14 (4), 439–462.

Eagleton, T. 1994[1991]. "Ideology and its Vicissitudes in Western Marxism." In *Mapping Ideology*, edited by S. Zizek, 179–226. London/New York: Verso.

Edelman, M. 1964. *The Symbolic Uses of Politics*. Urbana-Champaign: University of Illinois Press.

Eicher. 2007[2006]. *Eicher City Map: Delhi*. New Delhi: Eicher Goodearth Limited.

Ekers, M. and A. Loftus. 2008. "The Power of Water: Developing Dialogues Between Foucault and Gramsci." *Environment and Planning D: Society and Space*, 26, 698–718. doi:10.1068/d5907.

Escobar, A. 1999, February. "After Nature: Steps to an Antiessentialist Political Ecology." *Current Anthropology*, 40 (I), 1–30.

———. (n.d.). Retrieved from http://www.unc.edu/~aescobar/html/interests.htm#cuatro (accessed January 28, 2011).

Express News Service. 2006. "CAG Report Slams MLA Funds: 'Unused, Misused'." *The Indian Express*, March 20. Retrieved from http://cities.expressindia.com/fullstory.php?newsid=174520 (accessed September 12, 2006).

Fainstein, S.S. 2000, March. "New Directions in Planning Theory." *Urban Affairs Review*, 35 (4), 451–478.

———. 2005. "Planning Theory and the City." *Journal of Planning Education and Research*, 25 (2),121–130.

———. 2005[2004], September. "Cities and Diversity: Should We Want It? Can We Plan for It?" *Urban Affairs Review*, 41 (1), 3–19.

———. 2006, April 29. "Planning and the Just City." *Conference on Searching for the Just City*. New York: GSAPP, Columbia University.

———. 2010. *The Just City*. New York: Cornell University Press.

Fauconnier, I. 1999. "The Privatization of Residential Water Supply and Sanitation Services: Social Equity Issues in the California and International Contexts." *Berkeley Planning Journal*, 13 (1), 37–73.

Ferguson, J. and A. Gupta. 2002. "Spatializing States: Toward Ethnography of Neoliberal Governmentality." *American Ethnologist*, 29 (4), 981–1002.

Fontana, A. and M. Bertani. 2003. "Situating the Lectures." In *Society Must be Defended. Lectures at the Collège de France 1975–1976*, 273–293. New York: Picador.

Foucault, M. 1975[1972]. *Discipline and Punish: The Birth of the Prison*. Translated from French by A.M. Sheridan. London: Penguin/Peregrine.

———. 1978. "Governmentality." In *The Foucault Effect: Studies in Governmentality*, edited by G. Burchell, C. Gordon and P. Miller. Translated from French by R. Braidotti and revised by C. Gordon (published in 1991), 87–104. Chicago, IL: University of Chicago Press. [Also, in Faubion, J.D. ed., *Essential Works of Foucault 1954–1984: Power* (published in 2001, vol. 3), 201–222. London: Penguin.]

Foucault, M. 1984[1967]. "Of Other Spaces, Heterotopias." *Architecture/Mouvement/Continuité* (October 1984) 5, 46–49. Original lecture by Foucault, M. (March 1967) '*Des Espace Autres*'. English version: Foucault, M. (Spring 1986) "Of Other Spaces: Utopias and Heterotopias" (translated by Miskowiec J.), *Diacritics,*16 (1), 22–27.

———. 1984. *The Foucault Reader.* Edited by P. Rabinow. New York: Pantheon Books.

Friedmann, J. 2002. *The Prospect of Cities.* Minneapolis, MN: University of Minnesota Press.

Fuchs, C. 2003, April. "Structuration Theory and Self-organization." *Systemic Practice and Action Research*, 16 (2).

Gambhir, J.C. 1999. *Land as a Resource for Urban Development: Potentials and Limitations* (Research Report No. 20). As part of the Decentralized Training for Urban Development Project (An Indo-Dutch collaboration), Human Settlement Management Institute, New Delhi.

Gandy, M. 2004. "Rethinking Urban Metabolism: Water, Space and the Modern City." *City*, 8 (3), 363–379.

Ghertner, D.A. 2008. "An Analysis of New Legal Discourse Behind Delhi's Slum Demolitions." *Economic and Political Weekly*, 43 (20), 57–66.

Ghosh, A. and S. Tawa Lama-Rewal. 2005. *Democratization in Progress: Women and Local Politics in Urban India.* New Delhi: Tulika Books.

Giddens, A. 1982. *Profiles and Critiques in Social Theory.* Berkeley & Los Angeles: The University of California Press. Retrieved from http://books.google.co.in (accessed February 16, 2011).

———. 1984. *The Constitution of Society: Outline of the Theory of Structuration.* Cambridge: The Polity Press.

Goldstein, P. 2004, July. "Between Althusserian Science and Foucauldian Materialism: The Later Work of Pierre Macherey." *Rethinking Marxism*, 16 (3), 327–337.

González, M. 2004. "Postmodernism, Historical Materialism, and Chicana/o Cultural Studies." *Science and Society: A Journal of Marxist Thought and Analysis*, 68 (2, Summer), 161–186.

Government of NCT Delhi. 2001–2002. *Economic Survey of Delhi 2001–2002.* New Delhi: Planning Department. Retrieved from http://delhiplanning.nic.in/Economic%20Survey/Ecosur2001-02/Ecosur2001-02.htm (accessed June 30, 2009).

———. 2003. *Consolidated MLA Statement 2002–03.* New Delhi: Department of Urban Development.

———. 2004. *Economic Survey of Delhi 2003–2004.* New Delhi: Planning Department. Retrieved from http://delhiplanning.nic.in/Economic%20Survey/Ecosur2003-04/Ecosur2003-04.htm (accessed July 10, 2010).

Government of NCT Delhi. 2004–2005. *An Appraisal of Annual Plan 2004–05.* New Delhi: Planning Department.

Graham, S. and S. Marvin. 1996. *Telecommunications and the City: Electronic Spaces, Urban Places.* New York: Routledge.

———. 2001. *Splintering Urbanism: Networked Infrastructures, Technological Mobilities and the Urban Condition.* London: Routledge.

Grover, V. 2002. *From the Periphery to the Centre: A Rights-based Approach to Urban Poverty.* National Institute of Urban Affairs and Care-PLUS Report, New Delhi.

Gupta, S.S. 1992. *Goal and Strategies: Integrated Development Plan for India.* New Delhi: Concept Publishing Company. Retrieved from http://books. google.com/ (accessed May 15, 2010).

Habermas, J. 1995, March. "Reconciliation Through the Public Use of Reason: Remarks on John Rawls's Political Liberalism." *The Journal of Philosophy,* 92 (3), 109–131.

Hall, S. 1985, June. "Signification, Representation, Ideology: Althusser and the Post-structuralist Debates." *Critical Studies in Mass Communication,* 2 (2), 91–114.

Harvey, D. 1975[1973]. *Social Justice and the City.* London: Edward Arnold Ltd.

———. 1990. *The Condition of Postmodernity—An Enquiry into the Origins of Cultural Change.* Cambridge: Blackwell Publishers.

———. 1996. *Justice, Nature and Geography of Difference.* Cambridge: Blackwell Publishers.

———. 1997. "Contested Cities: Social Process and Spatial Form." In *The City Reader,* edited by R.T. LeGates and F. Stout (third edition, 2003), 235–243. London/New York: Routledge.

———. 2000. "Cosmopolitanism and the Banality of Geographical Evils." *Public Culture,* 12 (Spring, 2), 529–564.

———. 2003[1992]. "Social Justice, Postmodernism and the City." In *Designing cities: Critical readings in urban design,* edited by A.R. Cuthbert, 101–115. Malden, MA, USA :Blackwell Publishing.

Hazards Centre. 1999. *This City is Ours: Delhi's Master Plan and Delhi's People.* New Delhi: Report of the first Sajha Manch Convention, June 6, 1999. Delhi: Hazards Center.

———. 2004. *A Report on the Status of the First 28 Families Relocated from Hathi Ghat (Madanpur Khaddar,Yamuna Pushta).* Delhi: Hazards Centre.

Healey, P. 2003. "Collaborative Planning in Perspective." *Planning Theory,* 2 (2), 101–124.

Hetherington, K. 1997. *The Badlands of Modernity: Heterotopia and Social Ordering.* London: Routledge.

Hobsbawm, E. 2011. *How to Change the World: Tales of Marx and Marxism.* London: Little, Brown and Company.

Holston, J. 1989. *The Modernist City: An Anthropological Critique of Brasilia.* Chicago: University of Chicago Press.

Holston, J. 1998. "Spaces of Insurgent Citizenship." In *Making the Invisible Visible: A Multicultural Planning History*, edited by L. Sandercock, 37–56. Los Angeles, CA: University of California press.

———. 2008. *Insurgent Citizenship*. New York: Princeton University Press.

Holston, J. and A. Appadurai. 1996. *Cities and Citizenship: Public Culture*, 8 (2, Winter), 187–204.

Hoyt, L., R. Khosla, and C. Canepa. 2005. "Leaves, Pebbles, and Chalk: Building a Public Participation GIS in New Delhi, India." *Journal of Urban Technology*, 12 (1), 1–19.

IL&FS EcoSmart Ltd. 2006, October. *City Development Plan: Delhi*. New Delhi: JNNURM, Department of Urban Development, Government of Delhi.

Israel, G.D. 2009[1992]. *Determining Sample Size—PEOD6: Series of the Agricultural Education and Communication Department*. Florida Cooperative Extension Service, Institute of Food and Agricultural Sciences, University of Florida. Retrieved from http://edis.ifas.ufl.edu/pd006 (accessed March 10, 2010).

Jacobs, J. 1964[1961]. *The Death and Life of Great American Cities*. Harmondsworth: Penguin.

Jain, A.K. 1990. *The Making of a Metropolis: Planning and Growth of Delhi*. New Delhi: National Book Organisation.

———. 2009. *Urban Planning and Governance: A New Paradigm*. New Delhi: Bookwell.

Jessop, B. 1990. *State Theory: Putting Capitalist States in Their Place*. University Park, PA: The Pennsylvania State University Press.

———. 2007. "From Micro-powers to Governmentality: Foucault's Work on Statehood, State Formation, Statecraft and State Power." *Political Geography*, 26 (1), 34–40.

Jessop, B., N. Brenner, and M. Jones. 2008. "Theorizing Socio-spatial Relations." *Environment and Planning D: Society and Space*, 26 (3), 389–401.

JNNURM Primer. ca. 2005. *Property Tax: ULB Reform*. Retrieved from http://www.indiaurbanportal.in/JNNURM/mandatory_reforms/Local%20Level%20Reforms/PropertyTax.pdf (accessed August 25, 2010).

Kaika, M. 2005. *City of Flows: Modernity, Nature and the City*. London/New York: Routledge.

Kaviraj, S. 2005, April 8–10. *On the Enchantment of the State: Indian Thought on the Role of the State in the Narrative of Modernity*. New York: Columbia University. Retrieved from http://www.columbia.edu/cu/polisci/cspt/papers/2005/kaviraj05.pdf (accessed September 29, 2010).

———. 2010. *The Imaginary Institutions of India: Politics and Ideas*. New York: Columbia University Press.

Keil, R. 2005. "Progress Report—Urban Political Ecology." *Urban Geography*, 26 (7), 640–651.

Kelbaugh, D. and K.K. McCullough (eds). 2008. *Writing Urbanism: A Design Reader*. London/New York: Routledge.

Khilnani, S. 2001. "The Development of Civil Society." In *Civil Society: History and Possibilities*, edited by S. Kaviraj and S. Khilnani, 11–32. Cambridge, UK: Cambridge University Press.

Kleniewski, N. (ed.) 2005. *Cities and Society*. London: Blackwell Publishing Ltd.

Kooy, M. and K. Bakker. 2008, June. "Technologies of Government: Constituting Subjectivities, Spaces, and Infrastructures in Colonial and Contemporary Jakarta." *International Journal of Urban and Regional Research*, 32 (2), 375–391.

Kothari, R. (1991). "State and Statelessness in Our Time." *Economic and Political Weekly*, 26 (11 and 12), 553–558.

Krier, L. 1990. "Urban Components." In *New Classicism*, edited by L. Krier, 197–211. New York: Rizzoli.

Krumholz, N. and J. Forester. 1990. *Making Equity Planning Work: Leadership in the Public Sector*. Philadelphia, PA: Temple University Press.

Kumar, A. 2006. "Equitable Policies, Inequitable Outcomes and Inequitable Policies and Inequitable Outcomes: A Case of Land, Housing and Transport in Delhi." *ITPI Journal*, 3 (4), 1–14.

———. 2008. "Inclusive Planning and Development in the National Capital Territory of Delhi." *ITPI Journal*, 5 (4), 12–20.

———. 2009. "Capability Approach and the Distribution of Land, Housing and Access to Metro in the National Capital Territory of Delhi." In *Governance and Poverty Reduction: Beyond the Cage of Best Practices*, edited by A. Singh, K. Kapoor, and R. Bhattacharyya. New Delhi: PHI Learning Private Limited.

Kumar, G. 2005. "The Implementation of MP/MLALADS in Delhi and Its Overlap with the Imperatives of the 74th (Constitutional Amendment) Act." *Workshop on Urban Actors, Policies and Governance*, New Delhi, September 14. Retrieved from http://www.csh-delhi.com/UAPG/Publications/Workshops/Abstracts/Delhi/GK.htm (accessed September 12, 2006).

Kundu, A. 2006. *New Forms of Governance in Indian Mega-cities: Decentralisation, Financial Management and Partnerships in Urban Environmental Services*. Final report. A Study Sponsored under Indo-Dutch Programme for Alternate Development, NewDelhi and The Hague.

Lalchandani, N. 2009. "45% of Water is Lost in Transit." *The Times of India*, June 27, p. 4.

———. 2011. "Govt Looks at Private Firms to Solve Water Woes." *The Times of India*, March 10, p. 10.

———. 2013. "AAP Delivers on Water Promise, but Bills to Rise for Big Consumers." *The Times of India*, December 31. Retrieved from http://timesofindia.indiatimes.com/city/delhi/AAP-delivers-on-water-promise-but-bills-to-rise-for-big-consumers/articleshow/28155795.cms? (accessed January 10, 2014).

Lall, S.V. and U. Deichmann. 2006. *Fiscal and Distributional Implications of Property Tax Reforms in Indian Cities* (Working Paper No. 39). New Delhi: National Institute of Public Finance and Policy.

Landau, S. (ed.) 2002. *The New International Webster's Collegiate Dictionary of the English Language* (International Encyclopedic Edition, 2002 edition). Naples, Florida, USA: Trident Press International.

Larner, W. 2000. "Neo-liberalism: Policy, Ideology, Governmentality." *Studies in Political Economy*, 63. Retrieved from http://spe.library.utoronto. ca/index.php/spe/article/viewFile/6724/3723 (accessed September 22, 2010).

Lasswell, H. 1936. *Politics: Who Gets What, When, How.* New York: McGraw Hill.

Latour, B. 1996. "On Actor-network Theory: A Few Clarifications Plus More than a Few Complications (English Version)." *Soziale Welt*, 47, 369–381.

Lefebvre, H. 1991[1974]. *The Production of Space*. Translated from French by D. Nicholson-Smith. Oxford: Blackwell.

LeGates, R.T. and F. Stout (eds). 2003[1996]. *The City Reader* (3rd edition). London/New York: Routledge.

Legg, S. 2005. "Foucault's Population Geographies: Classifications, Bio-politics and Governmental Spaces." *Population, Space and Place*, 11, 137–156. Retrieved from http://onlinelibrary.wiley.com/doi/10.1002/psp.357/pdf (accessed September 22, 2010).

———. 2007. *RSG-IBG Book Series. Spaces of Colonialism: Delhi's Urban Governmentality*. Malden, MA, USA: Blackwell Publishing.

Lemke, T. 2001. "'The Birth of Bio-politics'—Michel Foucault's Lecture at the Collège de France on Neo-liberal Governmentality." *Economy and Society*, 30 (2), 190–207.

———. 2002. "Foucault, Governmentality, and Critique." *Rethinking Marxism*, 14 (3), 49–64.

———. 2007. "An Indigestible Meal? Foucault, Governmentality and State Theory." *Distinktion: Scandinavian Journal of Social Theory*, 8 (2), 43–64.

Lipietz, A. 1980. "The Structuration of Space, the Problem of Land and Spatial Policy." In *Regions in Crisis: New Perspectives in European Regional Theory*, edited by J. Carney, R. Hudson, and J. Lewis, 60–75. London: Croom Helm.

Logan, J.R. and T. Swanstrom. 2005[1990]. "Urban Restructuring: A Critical View." In *Cities and Society*, edited by N. Kleniewski, 28–42. London: Blackwell Publishing Ltd.

Lynch, K. 1960. *The Image of the City*. Cambridge, MA: MIT Press.

Maidan, M. 2008. "Foucault in the Geographer's Den." *Borderlands e-journal*, 7 (1) [Book Review of J.W. Crampton and S. Elden (eds). 2007. *Space, Knowledge and Power: Foucault and Geography*. Hampshire: Ashgate Publishing Ltd].

Mara, D. and G. Albaster. 2008. "A New Paradigm for Low-cost Urban Water Supplies and Sanitation in Developing Countries." *Water Policy*, 10 (2), 119–129.

Maria, A. 2008. "Urban Water Crisis in Delhi—Stakeholders Responses and Potential Scenarios of Evolution." *Iddri—Idées pour le débat*, 6 (June), 1–23.

Marques, E.C. and R.M. Bichir. 2003, December. "Public Policies, Political Cleavages and Urban Space: State Infrastructure Policies in Sao Paolo, Brazil, 1975–2000." *International Journal of Urban and Regional Research*, 27 (4), 811–827.

Marshall, T.H. 2009[1950]. "Citizenship and Social Class." In *Inequality and society*, edited by J. Menza and M. Sauder, 148–154. New York: W.W. Norton and Co. New York.

Mason, J. 2002. "The Ties that Bind: Infrastructure as the Defining Role of Planning." *Berkeley Planning Journal*, 16 (1), 77–87.

Mathur, O.P., D. Thakur, and N. Rajadhyaksha (with assistance from R. Bahl). 2009. *Urban Property Tax Potential in India*. New Delhi: National Institute of Public Finance and Policy.

MCD (Municipal Corporation of Delhi). 2005. *Civic Guide*, 15–19. New Delhi: Municipal Corporation of Delhi.

———. 2010. "Constituency-wise Expenditure and Physical Achievement under MLALAD Fund: Year 04–05, 05–06, 06–07 up to 28.02.10." *Progress Report of MLALAD Scheme*. Retrieved from http://www.mcdonline.gov. in/mcdengg/doclist.php?id=6 (accessed October 20, 2010).

McCluskey, W.J. (ed.). 1999. *Comparative Property Tax Systems: An International Comparative Review*. Aldershot, UK: Avebury Publishing Limited.

McFarlene, C. 2008, June. "Governing the Contaminated City: Infrastructure and Sanitation in Colonial and Post-colonial Bombay." *International Journal of Urban and Regional Research*, 32 (2), 415–435.

McLeod, M. 1996. "Everyday and 'Other' Spaces." *Architecture and Feminism*. Retrieved from http://web.mac.com/davidrifkind/fiu/library_files/mcleod.everyday-and-other-spaces.lib-iss.pdf (accessed March 11, 2010).

Mehta, P.B. 2011. "The Politics of Social Justice." *Business Standard India 2011*, 3–32. New Delhi: Business Standard Books.

Menon Sen, K. and G. Bhan. 2008. *Swept off the Map: Surviving Eviction and Resettlement in Delhi*. New Delhi: Yoda Press.

Miles, M., T. Hall, and I. Borden (eds). 2000. *The City Cultures Reader*. London/ New York: Routledge.

Miller, P. and N. Rose. 2008. *Governing the Present*. Cambridge: Polity Press.

Ministry of Environment and Forests: GoI and GNCTD. 2001, January. *Delhi Urban Environment and Infrastructure Improvement Project (DUIIP), Delhi 21*. New Delhi: Government of India and Government of National Capital Territory of Delhi.

Ministry of Statistics and Programme Implementation. 2010–2011. *Member of Parliament Local Area Development Scheme: Annual Report 2010–11.* New Delhi: Government of India.

Ministry of Urban Development. ca. 2008. *Handbook of Service Level Benchmarking.* New Delhi: Government of India.

Mitchell, W.J. 1999. *E-topia: Urban Life, Jim, but Not As We Know It.* Cambridge, MA: MIT Press.

Mollenkopf, J. 1992. "How to Study Urban Political Power." In *The City Reader*, edited by R.T. LeGates and F. Stout, 3rd edition, 235–243 (2003[1996]). London/New York: Routledge.

Molotch, H. 1976, September. "The City as a Growth Machine: Toward a Political Economy of Place." *The American Journal of Sociology*, 82 (2), 309–332.

———. 1993. "The Political Economy of Growth Machines." *Journal of Urban Affairs*, 15 (1), 29–53.

Montag, W. 1995. "'The Soul is the Prison of the Body': Althusser and Foucault, 1970–1975." *Yale French Studies*, 88, 53–77.

Municipal Valuation Committee-III. 2010, June. *Interim report*. New Delhi: Government of National Capital Territory of Delhi. Retrieved from http://www.mcdpropertytax.in/doclist/MVC-III.pdf (accessed February 10, 2014).

Musgrave, R. and P. Musgrave. 1984. *Public Finance in Theory and Practice.* New York: McGraw-Hill.

Nallathiga, R. 2008. "Metropolitan Urban Governance Approaches and Models: Some Implications for Indian Cities." Paper presented at *The Third International Conference on Public policy and Management on "Urban Governance and Public-Private Partnerships,"* IIM, Bangalore.

Naresh, G. and A.K. Halen. 2006, December 9. "Reforms in Property Taxation in India: Where Do We Stand? (Presentation)." International Seminar on *Reforms in Fiscal and Monetary Policies: The Road Ahead.* Foundation for Public Economics and Policy Research, New Delhi.

Navdanya/Research Foundation for Science, Technology and Ecology. 2005. *Financing Water Crises: World Bank, International Aid Agencies and Water Privatisation.* New Delhi: Research Foundation for Science, Technology and Ecology.

NCRPB (National Capital Region Planning Board). 1999. *Delhi-A Fact Sheet.* New Delhi: National Capital Region Planning Board.

Nehru, J. 1946[1994]. *The Discovery of India.* New Delhi: Jawaharlal Nehru Memorial Fund.

NIUA (National Institute of Urban Affairs). 2005. "Statistical Volume 1: Water Supply and Water Tariff 1999." *Status of Water Supply, Sanitation and Solid Waste Management of Urban India.* New Delhi: National Institute of Urban Affairs.

Nussbaum, M. 1993. *The Quality of Life.* Oxford: Clarendon Press.

O'Farrell, C. 2005. *Michel Foucault.* London/California/New Delhi: SAGE Publications.

O'Farrell, C. 2007. *michel-foucault.com: Key Concepts*. Retrieved from http://www.michel-foucault.com/concepts/ (accessed 30 September 2011).

Oldenburg, P. 1978[1976]. *Big City Government in India*. New Delhi: Manohar Publications.

Olssen, M. 2004. "Foucault and Marxism: Rewriting the Theory of Historical Materialism." *Policy Futures in Education*, 2 (3 and 4), 454–482.

PAC (Performance Audit Committee). 2006. *Performance audit of Strengthening and Augmentation Infrastructure Facilities in Assembly Constituencies in Delhi*. Report on Government of NCT of Delhi of 2006. Chapter-II. Comptroller and Auditor General of India (CAG). pp. 26-49. Retrieved from http://www.saiindia.gov.in/english/home/Our_Products/Audit_Report/Government_Wise/state_audit/recent_reports/Delhi/2005/Performance_Audit/Performance_Audit_Delhi_2005/civilvolII_chapter_2.pdf (accessed June 10, 2010).

Pahl, R.E. 1975[1970]. *Whose City? And Further Essays on Urban Society*, 2nd edition. Hammondsworth: Penguin.

Parivartan. ca. 2005. *Delhi Water Supply and Sewerage Project—An Analysis by Parivartan*. Retrieved from http://planningcommission.nic.in/data/ngo/csw/csw_5.pdf (accessed June 30, 2009).

Park, B.G. 2008. "Uneven Development, Inter-scalar Tensions, and the Politics of Decentralization in South Korea." *International Journal of Urban and Regional Research*, 32 (1), 40–59. doi:10.1111/j.1468-2427.2008.00765.x.

Parrott, M.R.M. 2002[1996]. *The Ethos of Modernity: Foucault and Enlightenment with the Historical Basis of Panopticism*. South Carolina: Rimric Press. Retrieved from www.mrmparrott.com (accessed November 14, 2004).

Pêcheux, M. 1994[1982]. "The Mechanism of Ideological (mis)Recognition." In *Mapping Ideology*, edited by S. Zizek, 141–151. London/New York: Verso.

Phillips, D.J.H., S. Attili, S . McCaffrey, and J.S. Murray. 2007. "Factors relating to the equitable distribution of water in Israel and Palestine." In *Water Resources in the Middle-East: The Israeli-Palestine Water Issues—From Conflict to Cooperation*, edited by H. Shuvel and H. Dweik, 249–256. Berlin: Springer.

Pinto, M.R. 2000. *Metropolitan City Governance in India*. New Delhi: SAGE Publications.

Pope, A. 2008, January–February. "Terminal Distribution." *Architectural Design*, 78 (1), 16–21 (Profile No. 191).

Prahalad, C.K. 2005. *The Fortune at the Bottom of the Pyramid: Eradicating Poverty through Profits*. Wharton School Publishing. Retrieved from http://books.google.co.in (accessed June 30, 2011).

PWC (Price Waterhouse Coopers), GHV, TCE. 2004. *Delhi Water Supply and Sewerage Project: Project Preparation Study* (DFR 3—Part B, Water Supply, Vol. I). Report prepared for Delhi Jal Board, New Delhi.

Retrieved from http://delhijalboard.nic.in/djbdocs/whats_new/news/pdf/
DFR3-Water%20Supply-Vol%20I-17%20Nov%202004.pdf (accessed
September 10, 2008).

PWC (Price Waterhouse Coopers), GHV, TCE. 2005. *Delhi Water Supply and
Sewerage Project: Project Preparation* (Final report, Part A, Vol. I).
Report prepared for Delhi Jal Board, New Delhi.

Rahman, N.A. 2013. "MLAs Sit on Local Area Funds." *The Times of India*,
March 22. Retrieved from http://timesofindia.indiatimes.com/city/delhi/
MLAs-sit-on-local-area-funds/articleshow/19117648.cms (accessed
October 20, 2013).

Rancière, J. 1999. *Disagreement: Politics and Philosophy*. Translated from
French by J. Rose. Minneapolis, MN: University of Minnesota Press.

Rao, P.S.N. 1993. "Private Sector Involvement in Urban Water Supply
Provision." In *Urban Studies Series: No. 1. Urbanization in Developing
Countries: Basic Services and Community Participation*, edited by Bidyut
Mohanty, 298–301. New Delhi: Institute of Social Sciences.

Rawls, J. 1969. "Distributive Justice." In *Philosophy, Politics and Society*, edited
by P. Laslett and W.G. Runciman, 3rd series. Blackwell.

———. 1971. *A Theory of Justice*. Cambridge, MA: Belknap Press (Harvard
University Press).

———. 1993. *Political Liberalism*. New York: Columbia University Press.

———. 1999. *Collected Papers*. Edited by S. Freeman. Cambridge, MA:
Harvard University Press.

Ribeiro, E.F.N. 2007. Community Consultation: A Critique of Chapter 15 of
the Delhi City Development Plan. *Delhi Citizen Critique of the City
Development Plan: Facilitating National Urban Renewal Mission in
Delhi*. Draft for discussion, compiled by the Centre for Civil Society for
a Seminar held on June 18, 2007 at India Habitat Centre, New Delhi, pp.
77–82.

Rogers, P. and A.W. Hall. 2003. *Effective Water Governance*. TEC Background
Paper No. 7. Sweden: Global Water Partnership Technical Committee.

Rose, N. 1999. *Powers of Freedom: Reframing Political Thought*. Cambridge:
Cambridge University Press.

Rowe, C. and F. Koetter. 1983[2001]. *Collage City*, 10th edition. London/Cambridge,
MA: The MIT Press.

Roy, A. 2003, March. "Paradigms of Propertied Citizenship: Transnational
Techniques of Analysis." *Urban Affairs Review*, 38 (4), 463–491. doi:
10.1177/1078087402250356.

———. 2005. "Urban Informality: Towards an Epistemology of Planning."
Journal of the American Planning Association, 71 (2), 147–158.

———. 2009a. "Why India Cannot Plan Its Cities: Informality, Insurgence
and the Idiom of Urbanization." *Planning Theory*, 8 (1), 76–87. doi:
10.1177/1473095208099299.

Roy, A. 2009b. "Civic Governmentality: The Politics of Inclusion in Beirut and Mumbai." *Antipode*, 41 (1), 159–179.

Roy, A. and N. AlSayyad (eds). 2003. *Urban Informality: Transnational Perspectives from the Middle East, Latin America, and South Asia.* Lanham: Lexington Books.

Roy, D. 2000. "Plan for the Masters." *The Hindustan Times*, November 14.

Ruet, J., V.S. Saravanan, and M-H. Zérah. 2002. *The Water and Sanitation Scenario in Indian Metropolitan Cities: Resources and Management in Delhi, Calcutta, Chennai, Mumbai* (CSH Occasional Paper No. 6). Publication of French Research Institutes in India

Runciman, W.G. 1966. *Relative Deprivation and Social Justice.* London: Routledge & Kegan Paul.

Sahoo, S. 2007. *Globalization and "the Politics of the Governed": Redefining Governance in Liberalized India.* Retrieved from http://sahoo.files. wordpress.com/2007/10/sahoo-working-paper-184.pdf (accessed October 23, 2010).

Saldanha, A. 2008. "Heterotopia and Structuralism." *Environment and Planning*, 40, 2080–2096. doi:10.1068/a39336.

Sandercock, L. 1997. *Towards Cosmopolis.* New York: John Wiley.

Sarkar, U. 2009. "Have Cash, Won't Spend, MCD's Councillors Seldom Use Rs 2cr They Get for Ward Development." *Delhi Scoop*, New Delhi, June 29. Retrieved from http://www.delhiscoop.com/story/2009/6/29/.../4283 (accessed July 3, 2010).

———. 2010. "Revision in Property Tax Slabs: Metro, Prices Trigger Tax Revamp in Delhi." *Delhi Scoop*, New Delhi, June 12. Retrieved from http://www.delhiscoop.com (accessed August 26, 2010).

Sassen, S. 2000[1994]. "The New Inequalities within Cities." In *The City Cultures Reader*, edited by M. Malcom, T. Hall, and I. Borden, 60–66. London: Routledge.

Savage, D. and S. Dasgupta. 2006. "Governance Framework for Delivery of Infrastructure Services." *India Infrastructure Report.* New Delhi: 3i Network/Oxford University Press.

Schirato, T. and J. Webb. 2003[2006]. *Understanding Globalization.* London/ California/New Delhi: SAGE Publications.

Scott, J.C. 1972, March. "Patron–Client Politics and Political Change in Southeast Asia." *The American Political Science Review*, 66 (1), 91–113. Retrieved from http://www.jstor.org/stable/1959280 (accessed September 27, 2009).

———. 1998. *Seeing Like a State: How Certain Schemes to Improve the Human Condition Have Failed.* New Haven/London: Yale University Press.

Selman, B. 2008, September. *Pirate Heterotopias.* Paper presented in part at the Failing Better, Goldsmith's First Annual Pan-MA Graduate Conference. Published in Deptford.TV.

Sen, A. 1979. "Equality of What?" *The Tanner Lecture on Human Values*. Lecture delivered at Stanford University, May 22. Retrieved from http://culturability.fondazioneunipolis.org/wp-content/blogs.dir/1/files_ mf/1270288635equalityofwhat.pdf (accessed September 2, 2011).

Sen, A. 2009. *The Idea of Justice*. London: Allan Lane.

Shah, P. and M. Bakore (eds). 2006. *Ward Power: Decentralized Urban Governance*. New Delhi: Centre for Civil Society.

Shane, D.G. 2007[2005]. *Recombinant Urbanism: Conceptual Modeling in Architecture, Urban Design and City Theory*. Chichester, England: John Wiley & Sons Limited.

Sharan, A. 2006. "In the City, Out of Place: Environment and Modernity, Delhi 1860s to 1960s." *Economic and Political Weekly*, 41 (47), 4905–4911.

Siddiqui, K., N. Ranjan, and S. Kapuria. 2004. "Delhi." In *Megacity Governance in South Asia—A Comparative Study*, edited by K. Siddiqui et al., 189–275. Dhaka: University Press Limited.

Simmel, G. 1950[1900]. "Metropolis and Mental Life." In *Sociology of Georg Simmel*, edited by K. Wolff. Glencoe, Illinois, USA: The Free Press. [Original source: Simmel, G. 1900. *The Philosophy of Money*.]

Singh, K.K. and S. Shukla. 2005. *Profiling "Informal City" of Delhi: Policies, Norms, Institutions and Scope of Intervention*. New Delhi: WaterAid India and Delhi Slum Dwellers Association.

Skocpol, T. 1985. "Bringing the State Back in: Strategies of Analysis in Current Research," In *Bringing the State Back in*, edited by P.B. Evans, D. Rueschemeyer and T. Skocpol, 3–42. Cambridge: Cambridge University Press.

Smith, N. 2008[1984]. *Uneven Development: Nature, Capital and the Production of Space*. Athens/London: The University of Georgia Press.

Soja, E. 2001. "Different Spaces: Interpreting the Spatial Organization of Societies." *Proceedings of the 3rd International Space Syntax Symposium, Atlanta*.

Sruthijith, K.K. 2003. "MLA Local Area Development Scheme." *Delhi Citizen Handbook 2003*. New Delhi: Centre for Civil Society. Retrieved from http://www.ccsindia. org/dh_pdf/ch_25mlalocalarea.pdf#search=%22MLA%20Local%20Area%20 Development%20Scheme%22 (accessed September 12, 2006).

Stavrides, S. 2007. "Heterotopias and the Experience of Porous Urban Space." In *Loose Space: Possibility and Diversity in Urban Life*, edited by K. Franck and Q. Stevens, 174–192. London: Routledge.

Stein, J.M. (ed.) 1995. *Classic Readings in Urban Planning*. New York: McGraw Hill.

Sundaram, R. 2010. *Pirate Modernity: Delhi's Media Urbanism*. New Delhi: Routledge.

SSAA (Sushant School of Art and Architecture). 2004. *Urban Morphology of Gurgaon* (unpublished report prepared by the fourth year students in the Urban Housing Studio 2004). Gurgaon: Sushant School of Art and Architecture.

————. 2011. *Morphological Study of Gurgaon* (unpublished report prepared by the fourth year students in the Urban Design Studio 2011). Gurgaon: Sushant School of Art and Architecture.

Susheela, A., M. Bhatnagar, and A. Kumar. 1996. *Status of Drinking Water in the Mega City—Delhi.* Paper presented at the 22nd WEDC Conference, New Delhi, India.

Swyngedouw, E. 1999. "Modernity and Hybridity: Nature, Regeneracionismo, and the Production of the Spanish Waterscape, 1890–1930." *Annals of the Association of American Geographers,* 89 (3), 443–465.

————. 2004. *Social Power and the Urbanization of Water—Flows of Power.* Oxford: Oxford University Press.

————. 2005, October. "Governance Innovation and the Citizen: The Janus Face of Governance-Beyond-the-State." *Urban Studies,* 42 (11), 1991–2006.

————. 2006. "Circulation and Metabolisms: (Hybrid) Natures and (Cyborg) Cities." *Science as Culture,* 15 (2), 105–121.

————. 2009, September. "The Antinomies of the Postpolitical City: In Search of a Democratic Politics of Environmental Production." *International Journal of Urban and Regional Research,* 33 (3), 601–620.

Tawa Lama-Rewal, S. 2005. *Health Services as an Analyzer of Urban Governance: A Study of Delhi* (based on the paper: "Urban Governance through the Prism of Primary Health Care Provision: A Study of Delhi"). Paper presented at the workshop on Urban Actors, Policies and Governance in Delhi, Jawaharlal Nehru University, NewDelhi, September 14.

Taylor, N. 1998. *Urban Planning Theory since 1948.* London: SAGE Publications.

Teyssot, G. 1977. "Heterotopia and History of Spaces." In *Architectural Theory since 1968,* edited by K.M. Hays (2000). Cambridge: MIT Press.

Thapar, R. 1987. "Dall'indipendenza ad oggi (What Has Happened Over the Years of Freedom)." *Space and Society,* 38 (April–June).

The Economic Times. 2010. "Panel Wants More Delhi Colonies in High Property Tax Slot." New Delhi, June 12. Retrieved from http://articles.economictimes.indiatimes.com/2010-06-12/news/27606118_1_municipal-valuation-committee-tax-rates-property-tax (accessed August 27, 2010).

The High Court of Delhi. 2010, January 14. *Siri Fort Road Residents Welfare Association and Mr. C.P. Sabharwal (petitioners) vs. Municipal Corporation of Delhi* (WPC No. 4905/2007). Order by Hon'ble Justice Mr Sanjiv Khanna, New Delhi.

The Hindu. 2009. *MCD Finalizes Budget with Additional Allocation,* January 17. Retrieved from http://www.thehindu.com/2009/01/17/stories/2009011759250300.htm (accessed March 7, 2010).

————. 2010. *Report on Property Tax Tabled,* July 29. Retrieved from http://www.hindu.com/2010/07/29/stories/2010072963560300.htm (accessed August 25, 2010).

The Indo-American Arts Council. 2011. *Juggad Urbanism.* Retrieved from http://www.iaac.us (accessed July 20, 2011).

The Urban Institute. 2007. *A New Land Title Registration System for Delhi: Recommendations.* Report prepared for the U.S. Agency for International Development/India, Washington, D.C.

The World Bank. 2005. *Project Information Document (PID): Concept Stage: (Project name) Delhi Water Supply and Sewerage Project.* (Project prepared on February 15, 2005). Retrieved from http://siteresources. worldbank.org/INTINDIA/Resources/DJB_PID_FOR_WEB.PDF (accessed September 14, 2009).

Times News Network. 2004. "Many Areas in City Go Waterless." *The Times of India*, New Delhi, May 31.

———. 2010. "Floor Area Ratio May Rise in Places near Metro, BRT." *The Times of India*, New Delhi, August 10, pp. 1–2.

———. 2011a. "Circle Rates Doubled in City, to Take Effect from Today." *The Times of India*, February 8, p. 2.

———. 2011b. "Delhi Govt Hikes MLALAD Fund to ₹4 Crore." *The Times of India*, June 15. Retrieved from http://articles.timesofindia.indiatimes. com/2011-06-15/delhi/29660991_1_delhi-govt-hikes-sheila-dikshit-fund (accessed October 10, 2011).

Titmus, R.M. 1962. *Income Distribution and Social Change.* George Allen and Unwin: London.

Tiwari, G. 2003. "Transport and Land Use Policies in Delhi." *Bulletin of the World Health Organization*, 81 (6), 444–450.

Tiwari, S. (ed.) 2005. *Education in India*, vol 1. New Delhi: Atlantic Publishers and Distributors.

Urbinati, N. 2003. "Can Cosmopolitan Democracy be Democratic?" In *Debating Cosmopolitics*, edited by D. Archibugi, 67–85. London: Verso.

Vedeld, T. and A. Sriddham. 2002. "Livelihoods and Collective Action among Slum-dwellers in a Mega-city, New Delhi." Retrieved from http://dlc.dlib.indianaedu/ archive/00000938/00/vedldt120402.pdf (accessed September 10, 2005).

Veenhoven, R. 2000. "Wellbeing in the Welfare State: Level Not Higher, Distribution Not More Equitable." *Journal of Comparative Policy Analysis*, 2, 91–125.

Verma, G.D. 2002. *Slumming India, a Chronicle of Slums and Their Saviors.* New Delhi: Penguin Books.

WBPCC (West Bengal Pradesh Congress Committee). 1995. "Congress Resolution." *Avadhi Session.* Retrieved from http://www.wbpcc.org/ resolution.htm (accessed September 9, 2011).

Willensky, H.L. and C.N. Lebeaux. 1965. *Industrial Society and Social Welfare.* New York: Free Press.

WMO (World Meteorological Organization). 1992. *International Conference on Water and the Environment: Development Issues for the 21st Century: The Dublin Statement and Report of the Conference.* Hydrology and Water Resources Department, WMO, Geneva.

Yamane, T. 1967. *Statistics: An Introductory Analysis*, 2nd edition. New York: Harper and Collins.

Young, I.M. 2011[1990]. *Justice and the Politics of Difference.* Princeton, New Jersey, USA: Princeton University Press.

Zérah, M.H. 2000a, September. "Household Strategies for Coping with Unreliable Water Supplies: The Case of Delhi." *Habitat International*, 24 (3), 295–307.

————. 2000b. *Water: Unreliable Supply in Delhi.* New Delhi: Manohar Publications.

Zizek, S. (ed.) 1994. *Mapping Ideology.* London/New York: Verso.

Index

About the Author

Suptendu P. Biswas is an architect, urban designer, and planning professional involved in teaching, research, and consultancy. He is a graduate from B.E. College (IIEST), Shibpur, and is a postgraduate and doctorate from the School of Planning and Architecture (SPA), New Delhi. He works on built environment, urbanism, spatial equity, and sociology of culture.

Dr Biswas is a recipient of the National Scholarship from Government of India and Senior Fellowship from Ministry of Culture, Government of India.

He is one of the partners of VSPB Associates, an Architecture, Urban Design, and Landscape firm in New Delhi, as well as a Visiting Faculty in postgraduate planning courses at SPA, New Delhi. He is also one of the founder trustees for the Trust for Sustainable Education and Action in Architecture (SEARCH), based at Kolkata and New Delhi.

Dr Biswas has taught in top institutions in Delhi NCR for one and a half decade including SPA, Sushant School of Art & Architecture, and GGS Indraprastha University. He also has a number of significant publications in journals and newspapers on architectural and urban issues, and has made presentations at national and international symposia too. In a professional career spanning more than two decades, his projects have been presented, published, and exhibited in India and abroad.